MIMESIS
INTERNATIONAL

ATMOSPHERIC SPACES
n. 19

What is an "Atmosphere"?

According to an aesthetic, phenomenological and ontological view, such a notion can be understood as a sensorial and affective quality widespread in space. It is the particular tone that determines the way one experiences her surroundings.

Air, ambiance, aura, climate, environment, genius loci, milieu, mood, numinous, lived space, Stimmung, but also Umwelt, ki, aida, Zwischen, in-between – all these words are names hiding, in fact, the founding idea of atmospheres: a vague ens or power, without visible and discrete boundaries, which we find around us and, resonating in our lived body, even involves us.

Studying atmospheres means, thus, a parte subjecti, to analyse (above all) the range of unintentional or involuntary experiences and, in particular, those experiences which emotionally "tonalise" our everyday life. A parte objecti, it means however to learn how atmospheres are intentionally (e.g. artistically, politically, socially, etc.) produced and how we can critically evaluate them, thus avoiding being easily manipulated by such feelings.

Atmospheric Spaces is a new book series whose aim is to become a point of reference for a community that works together on this philosophical and transdisciplinary subject and for all those whose research, more broadly, is involved in the so-called "affective turn" of the Social Sciences and Humanities.

Ola Sigurdson

THRESHOLDS AND PLACES

On Existential Spatiality 2

MIMESIS
INTERNATIONAL

www.mimesisinternational.com
e-mail: info@mimesisinternational.com

Book series: *Atmospheric Spaces*, n. 19

Series Issn: 3103-4322

Isbn: 9788869775567

P.I. C.F. 0241937030

CONTENTS

Prelude

Even if spatial experience by necessity shapes human existence, not all such experiences are alike. Of all the myriads of different experiences of spatiality possible, a fundamental distinction could be made, or so I would suggest, between moments of transition and sites of belonging. In everyday parlance, these are what we call thresholds and places. Both thresholds and places are realities of physical space but at the same time also existential structures. They simultaneously form our experience of spatiality on all levels of existence.

Thresholds and places, places and thresholds. Which one comes first? Can we do without the one or the other – continuously transitioning or just remaining where we belong? I would suggest not. Never arriving can be as fearful or as harmful as never being able to leave. As I will argue in this book, experiences of thresholds and of places are distinct, but nevertheless inseparable. Human existence unfolds in the dialectical tension between transitioning and belonging. To paraphrase Gillian Rose, a life without thresholds would be as mad as one without places.[1] Overlooking either thresholds or places could have fateful consequences for how we understand and live our life together.

This volume continues the exploration of existential spatiality that I began in *Atmospheres in Everyday Life*, turning attention to the dialectic between thresholds and places as core spatial structures of human experience. Thresholds are where we cross into the ambiguous and the unknown; places are where we find orientation and familiarity.

---

1 Rose 1995, p. 98.

Thresholds are approached as liminal zones – existential transitions marked by ambiguity and becoming – while places are studied as sites of stability, memory, and affective attachment. Together, these two dimensions reveal the lived complexity of space as something that is always both structured and structuring – never static, always becoming. Where *Atmospheres* explored the sensory and affective textures of everyday environments, *Thresholds and Places* turns to the spatial dynamics of transformation and belonging.

I have chosen the simplest structure possible for this book: first thresholds, then places. Even though inseparable, thresholds and places, both as phenomena and as concepts, need to be treated separately, not just for the sake of clarity but also because of a certain contemporary forgetfulness or even repression of both threshold and place, which calls for a distinct articulation of how both of them are encountered, experienced, and practiced in everyday life. As in *Atmospheres*, I shall draw on anthropology, architecture, literature, philosophy, and lived life, to show how we are always situated both between and within thresholds and places. First, then, thresholds. After that, places.

Part 1

Thresholds

At the Threshold

There are thresholds in every house. The most important ones lead into or out of the house. They are where the home and the outside world meet. Outside the house, in the city, there are also many thresholds, often marked concretely and materially. They separate one place from another place, one neighbourhood from another neighbourhood and one atmosphere from another atmosphere. Thresholds are everywhere. Yet there is not much talk about them. What are they and how do they affect us?

"The threshold" – *die Schwelle* (German), *le seuil* (French), *la soglia* (Italian), *tröskeln* (Swedish), *terskeln* (Norwegian), *ho oudos* (Greek), *limen* (Latin) – is a term for a thing or function that is everywhere but rarely the subject of any systematic investigation. Despite its ubiquity, most of the time we seem to cross it without much thought. Or do we? Is it that the historical customs that developed around thresholds – carrying the bride across the threshold, not shaking hands across a threshold, not crossing the threshold before being invited – are in some way still alive in our own time?[1] Do these indicate that we actually attach greater importance to thresholds than our relative silence about them suggests? The simple design of thresholds can hardly be the reason for this oversight. In another economy, the simple slab that forms the threshold of my office might be made of solid oak, but its artlessness does not prevent it from fulfilling its function. After all, my simple slab is also a threshold that makes colleagues or students hesitate a little to cross it without an explicit invitation. Is the elusive nature of thresholds therefore rather

1 A rich selection of historical customs is presented by Trumbull 1896.

a consequence of their very function? They may be a "demarcation on the floor between rooms" according to a Swedish dictionary, but experience teaches us that this demarcation carries a deep symbolic meaning, a meaning that structures our relationship to the physical spaces we inhabit.

As a compound term, we encounter "threshold" everywhere. By threshold states, we mean countries that have the capability to produce nuclear weapons or have secretly done so, as opposed to official nuclear weapon states and non-nuclear weapon states. When we talk about thresholds, we are talking about a certain intensity or magnitude being exceeded so that something occurs or becomes visible. With threshold training, we aim to improve our endurance by running as fast as we can, but not faster than the body can still transport the lactic acid away from the muscles. In these terms drawn from international relations, biology and exercise, thresholds refer to an intermediate state or a turning point, a peripety of a situation or event. Is it perhaps the instability and dynamic mobility implied here that makes them so elusive?

The thresholds between the home and the street, as well as between different neighbourhoods, are as much "demarcations between rooms" in the form of a concrete materiality of some kind, even if they do not run along a floor and consist of a crossbeam. The threshold, according to philosopher Otto Friedrich Bollnow, denotes "the border between inside and outside", but you can be "inside" and "outside" in relation to many different spaces.[2] You are at home in your house in a way that I am not; someone feels like a stranger in the Gothenburg neighbourhood Masthugget but not in the neighbourhood Gårda. It is precisely such border crossings between different spaces that define who we are, says Bollnow. Although not all thresholds are of the same dignity, the crossing of thresholds between different spaces can sometimes have "a deeper symbolic meaning: it is the place of transition to a new life".[3] It is precisely the relationship between thresholds as physically and geographically

2 Bollnow 2011, p. 150.
3 Bollnow 2011, p. 150.

concrete "demarcations between rooms" ("the border between inside and outside") and their symbolic meaning as "the place of transition to a new life" (an intermediate state or turning point) that I want to explore in the first part of this book.

My thesis is that thresholds are dynamic and elusive phenomena that in an existential sense structure how we inhabit the spaces we find ourselves in. Before setting out a direction for my investigation of thresholds as a phenomenon, I will substantiate and deepen this claim with three examples of how they have been treated literarily, religiously and philosophically. If it is true that there is a relative silence about thresholds, we first need to see them before we can understand what they are, how they work and how they affect us. Although none of the examples I give really have thresholds as their main subject, they say enough for us to recognise their existential significance.

The Vampire at the Threshold

Why does the threshold recur in representations of vampires? A vampire, if the popular culture narrative is to be believed, cannot cross the threshold of a home unless invited by the occupant. No vampire enters the home as an uninvited guest. There must be some consent on the part of the resident. In Bram Stoker's 1897 novel *Dracula*, a classic of the vampire genre and a major trendsetter, we encounter the threshold a number of times, not as a central phenomenon, but with sufficient concreteness to tell us something about its significance.

Dracula is an epistolary novel, in which we are confronted with Count Dracula mainly through diaries, letters and newspaper articles, and more rarely through a direct account. We first meet the lawyer Jonathan Harker, who, on behalf of a British law firm, visits the Count in Transylvania to assist with his imminent move to England. When he arrives at the Count's castle in the Carpathian Mountains, he is met at the gate by Dracula himself, who greets him with the following words: "Welcome to my house! Enter freely and of your own will!"[4] The Count, however, remains standing and does not go to meet Harker until Harker has stepped "over the threshold".[5] Here, we are at the very beginning of the story and cannot yet harbour many ideas about the significance of the threshold. After the opening scenes in which Harker will become Count Dracula's prisoner in the castle in Transylvania, the novel shifts to portraying the Count indirectly, through other characters' accounts of rumours or experiences of him. Yet we get another hint of the significance of the threshold for the vampire when the Amsterdam doctor and professor Abraham Van Helsing eventually figures out what kind of creature they are dealing with. He seals the entrance to the burial chamber of one of the Count's victims with a dough made from communion bread.[6] Only the consecrated bread can prevent the undead from returning to the tomb and its grave.

In the latter part of the novel Van Helsing gives a fuller account of the nature of the vampire – his vast and supernatural powers, but also his limitations – which we have hitherto met only in passing. The Count's power diminishes at dawn; he must be able to take refuge in his coffin and can only change his form at dinner or just at sunset or sunrise. Van Helsing's account makes explicit the limitations the vampire faces at the threshold: "He may not enter anywhere at the first, unless there be some one of the household who bid him to come; though afterwards he can come as he please."[7] A central part of the complication in *Dracula* is how the Count cunningly manages to overcome his limitations, for example by secretly inviting himself to the hospital where Van Helsing and the other vampire hunters meet, so that he can spy on them. This would not have been possible if one of the inmates, the mental patient Renfield, had not opened the window and called out to him: "Come in, Lord and Master!"[8] This is a knowledge that Count Dracula has to acquire through his own

4 Stoker 2011, p. 22.
5 Stoker 2011, p. 22.
6 Stoker 2011, pp. 224–227.
7 Stoker 2011, p. 255.
8 Stoker 2011, p. 299, cf. p. 322.

experience, but he is unable to change the very conditions imposed by his limitations. In *Dracula*, the importance of thresholds for the vampire is depicted both through the experiences of Dracula at the threshold and the conclusions that Van Helsing draws from these experiences.

The references to thresholds are not extensive in *Dracula* and they never really answer the question of why Count Dracula cannot cross a threshold without someone inviting him in. But the novel does give enough hints to allow us to speculate a little.

Dracula is a novel about transgressing boundaries and about intermediate states where boundaries are blurred: specifically between the living and the dead in the form of the undead, but also, in one interpretation of the novel, between the sexes. After Freud, it is clear that the vampire's hunger for blood carries sexual connotations.[9] And Count Dracula, like all vampires, is a shape-shifter, someone who is able to change his physical form. In the novel, the Count is portrayed in a variety of ways, both in terms of his physical constitution and appearance, but remains recognisable. So if the novel is about transitions, it is not surprising that the threshold is important because, in its concrete, physical form, it symbolises the intermediate state that exists in every transition. This is also the case in our own everyday lives: on the threshold I am neither in nor out, home nor away. On the threshold of my home, you are not yet at home with me, but on the other hand you are no longer out on the street. You are about to enter, and if you are not about to break in unlawfully, it is I or someone else in the household who has invited you in.

Even today, the threshold carries a symbolic meaning: here I am no longer who I was, but not yet who I will become. The fact that Count Dracula has to be invited to enter a house but can then move freely across the threshold suggests that everyone in the novel can, in a way, find themselves in the same position as the hospitalised patient Renfield in relation to Dracula, namely to willingly welcome the Count. Is there a deeper connection between the vampire and the

9 See de Lauretis 2008, pp. 26–32.

threshold? That the threshold plays an important role in stories about vampires may be merely a convention of such stories – a convention that Stoker helped to establish through his *Dracula* – but if so, it is a convention deeply rooted in the symbolic meaning of the threshold. The invitation to cross the threshold is – even outside the domain of the vampire novel – not only about a purely instrumental practicality, to enter a room, but most deeply about our desire: who do we want to meet, who is this other, who do we want to be and who do we become in relation to this other?

Existential Thresholds

The symbolic meaning of the threshold is ancient. In Ovid's calendar, the *Fasti*, Janus is the god of beginnings, gates, and transitions, while Terminus is the god of boundary markers. Janus is the "guardian of the threshold" and thus regulates the access of other gods, while Terminus sets the boundaries "of peoples, cities and vast kingdoms" without which war would be a fact.[10] It is a matter of keeping good relationships with these gods through sacrifice and remembrance, of respecting the threshold.

The existential and religious importance attached to thresholds throughout history is interestingly illustrated by a quote from the Neoplatonic philosopher Porphyry from the end of the second century. In an exegesis of a few lines from Homer's *Odyssey* about the cave of the nymphs, Porphyry says that Homer is aware that "gates are sacred" and that wise men of the Pythagoreans and Egyptians therefore forbade anyone to speak when passing through a gate or entrance.[11] The "gates" referred to – in Greek *thyra* with the synonym *pyle* (here closest to "entrance") – denote the very things called gates, but the context shows that it is the gate as a passage or passageway that is of concern to Porphyry here. It is therefore not inappropriate to translate *thyra*, as has sometimes been done,

10 Ovid 2011, 1,173, 2,659.
11 Porphyry 1986, p. 72 (§ 27). Cf. Homer 2016, book 13, lines 102–112.

as "threshold"; as a passage, the gate or entryway is a threshold between two rooms.[12] The quality of holiness attributed to the gate in its aspect of passage is denoted by the adjective *hierá*. Greek has several words for sacred, and *hierá* can be compared to *hagia*, the latter term referring to a quality of a person or a relationship with a person, while the former refers to an object, place or event. It is thus consistent that *he thyra* is *hierá* rather than *hagia*; the gate in its aspect of passage is at once an object, a place and an event.

There are only two gates to the cave. The Homeric gods, in the account of the cave of the Naiads commented on by Porphyry, enter through the southern gate but the northern gate, intended for humans, is also sacred. The silence imposed when someone crosses the threshold is an expression of respect for the god "who is the beginning of all things". The cave of the Naiads symbolises, in Porphyry's interpretation, the cosmos, and the gates of the cave therefore become symbols of a human being's relation to its own origin and to its insight into that origin, both its material creation and its liberation from it in the Neoplatonic doctrine of salvation.[13] In other words, the thresholds of the cave are a kind of concrete manifestation of a mystagogy, the initiation by which a human being is both concretely and figuratively brought into the mystery. This mystagogy is the ultimate subject of Porphyry's allegorical exposition of Homer's description of the cave of the Naiads in the *Odyssey*. According to Porphyry, Homer and Plato conveyed the same message, albeit in different forms. The cave of the Naiads with its southern and northern gates is a place of transition, of a fundamental change in human existential conditions.

For Porphyry, the liberation sought is about the liberation of the soul from the material world, but what I want to draw attention to here is that this spiritual escape is nevertheless symbolised by the gates as earthly passages. According to Porphyry, these gates

---

12 Gaston Bachelard quotes a French translation in Bachelard 2020, p. 306, which uses the noun "seuil" as a translation of *thyra*. The English translation (Bachelard 1994, p. 223) uses, accordingly, "threshold".

13 See Akçay 2019.

actually existed in the material world. He imagined that the cave of the Naiads existed as a concrete, geographically definable location, specifically on the Greek island of Ithaca. The symbolic meaning of the cave did not mean that its concrete existence was indifferent. Speaking of the cave and the gates as symbols is consistent not only with our own language but also with Porphyry's own: *symbolon* is a central term in his *De antro nympharum*. The concrete meaning of the symbol conveys a different, more abstract or profound meaning without itself being cancelled out.[14] The actual gate into and out of the cave is, at the same time, the threshold between earthly and heavenly life. It is therefore not surprising that the concrete manifestation of this existential passage is considered sacred by Porphyry, since these two dimensions are interrelated. The sacredness of the passage is not only a matter of its associations with divine liberation; passage across the threshold is also fraught with danger, since it is the very foundations of human existence that are literally at stake: a person is in the process of becoming, in their liberation from the material world, another, someone new. The passage through the gates of the cave is an existential turning point.

Gates, entrances, passages, all have a central role in ancient religions as symbols of a more existential passage. They are, in Bollnow's words quoted above, "the place of transition to a new life". Even when the threshold as such is not mentioned, they have a threshold function: the threshold as a transition from one state to another. The fact that the cave of the Naiads is perceived by the Neoplatonist Porphyry as a concrete place on Ithaca and not just as an image for an existential realisation is typical of the general way in which existential passages were thought of at this time as being linked to physical passages. Different philosophies or theologies in history thought differently about a place's meaning and the meaning of place. The meaning of Jerusalem as a holy city came to change for the early Christian Church in comparison to its Jewish origins through the Church's mission "to all nations" (Matthew 28:19). When the Temple in Jerusalem was demolished in 70 CE, Judaism

14 Akçay 2019, pp. 34 f.; Ricœur 1976, pp. 53–63.

was transformed into a diaspora religion. The constant is the place as a symbol, that is, the linking of geographical and existential meaning with the passage as threshold or the threshold as passage.

Such symbolic meanings are not unknown in our own time, in political and religious movements but also beyond them. As the historian of religion Mircea Eliade writes in a discussion about the ritual function of the threshold for sacred as well as profane space: "The threshold, the door *show* the solution of continuity in space immediately and concretely; hence their great religious importance, for they are symbols and at the same time vehicles of *passage* from one space to the other."[15] But he also traces a similar transition in the threshold of our home. The thresholds into and out of our homes are yet another reminiscence of a more sacred transition and are therefore also subject to special rites. And in a discussion on the dialectic between outside and inside, the philosopher Gaston Bachelard argues that the door is "an entire cosmos of the Half-open", one of the "primal images" as a symbol of the closed and the open as existential modes.[16] Even in our time, poetry can help us to sense "that, incarnated in the door, there is a little threshold god".[17] Admittedly, Bachelard refers such images to daydreaming and poetry, and therefore seems to largely separate concrete locality and existential meaning: the door is for him an image, not a symbol. But nevertheless, especially in his concrete elaborations of the meaning of space, a symbolic connection between locality and meaning is implied as the very premise of these poetic dreams – it is, after all, an original image.

It is this relationship between the concrete locality and the existential meaning of the threshold that I want to explore in this book. What does it look like? Has it been lost in modern times? Is it possible that the articulation of the existential meaning of the concrete thresholds has disappeared in our time or been referred to the subjective, more than to the experience of the threshold as a symbol?

---

15 Eliade 1959, p. 25.
16 Bachelard 1994, p. 222.
17 Bachelard 1994, p. 223.

A loss of language then does not simply mean that the threshold experience is impoverished, but that it has become homeless even though it has not disappeared completely.

The Threshold Between Inside and Outside

The fact that the threshold is a material thing with a concrete locality and therefore appears in the most everyday contexts has not escaped the attention of ancient philosophy. A simple definition of what is called *oudos* in Greek is found in Aristotle's *Metaphysics*: "a piece of wood or stone placed in such-and-such a way".[18] For Aristotle, the threshold is clearly an everyday phenomenon, as we often encounter it. This is why we can use his definition to further deepen our understanding of the threshold as a phenomenon. According to Aristotle, what defines a threshold is its position: that it is *between* something and something else, for example the entrance to a house, where it separates inside and outside. Aristotle is well aware of this property of the threshold, but chooses to argue that the intermediate state of the threshold is a consequence of its being between inside and outside, rather than the difference between inside and outside being a consequence of the threshold itself: "if the threshold [*ho oudos*] is defined by its position [*thesei*], the position is not derived from the threshold, but rather *vice versa*".[19]

Contra Aristotle, the intermediate state of the threshold is characterised both by the fact that it is established by the difference between inside and outside and that it establishes the difference between inside and outside itself. However, if Eliade writes about the difference between sacred and profane space, it is equally applicable to the difference between inside and outside in a more general sense: "The threshold is the limit, the boundary, the frontier that distinguishes and opposes two worlds – and at the same time

18 Aristotle 1933, VIII, 1043a.

19 Aristotle 1933, 1043b. But it is not quite as clear-cut as Aristotle claims, see Carter 2010.

the paradoxical place where these two worlds communicate".[20] Put another way, the threshold is both *copulative* and *privative*: it both unites and separates. Precisely by establishing a transition between inside and outside, the position of the threshold is produced as between two states and therefore also produces these two states that were not there as such before the difference between them was established. The function of the threshold is thus to be between inside and outside, thereby establishing the difference between inside and outside. The threshold not only produces differences but also connects the states that are separated. There is no inside except in relation to an outside. The threshold concept is a relational concept that is both copulative and privative.

As an intermediate state, a threshold is neither inside nor outside, but a transition between them. Whoever is on the threshold on the way out is no longer inside but not yet outside. On the threshold, we are no longer in a place we can call here – as in "here is my home" – but not yet in another place we can call there – as in "there is the street outside my home". We are neither here nor there. Of course we can say that it is here on the threshold that we are, but if we are talking about a threshold that is truly a transitional state and cannot be limited to a physical object – "wood or stone in such and such a position" – then this "here" is dynamic rather than static and does not denote a permanent or unchanging position. Rather, it is precisely the instability, ambiguity or transition that characterises the threshold. In other words, a threshold is neither identical with one side nor the other of that which it both unites and separates, nor is it a third thing that exists independently of that which it both unites and separates. Its mode of existence is different from the more stable and bounded states of an inside or an outside. It is not really a place in its own right, but an intermediate state between two places – a "state" that is dynamic rather than static.

There may indeed be significant differences between physical thresholds and experiential thresholds. While there may be both similarities and differences, it is essential to recognise that there

20 Eliade 1959, p. 25.

may well be fairly insignificant thresholds. Between my office at the university and the corridor outside, there is, as I mentioned, a simple threshold slab. During the course of a working day, I cross it many times without particularly noticing it, but when a colleague or student crosses the threshold on their way into my office, they can sometimes be seen hovering for a moment on the threshold, waiting for a concrete invitation. For a short question or discussion, stopping on the threshold may be enough, but a more in-depth dialogue requires the person to come in and perhaps sit down. For particularly sensitive conversations, we close the door. The concrete threshold plate is thus a symbol, a physical manifestation of an experiential difference between meeting in the corridor, in the office or at the threshold. The corridor and the office differ in terms of atmosphere in a way that makes a transition between the two necessary. This transition is symbolised by the physical threshold and is therefore also in a sense staged by it. On certain occasions, such as when I put a letter in the post box or fetch a cup of coffee, the physical threshold is, as I have said, of little importance to me, but it retains its symbolic significance for the colleague or student who finds the door open but the office empty and therefore hesitates to enter. When I myself put the key in the lock and accidentally open the door to a colleague's neighbouring office, I abruptly slow down my previously rapid steps: I am not invited here, I do not belong here. To step over the threshold would be an intrusion. At the threshold is a boundary.

Is the threshold therefore equal to a boundary, as Aristotle seems to think? In a sense it is, because it demarcates inside from outside. But unlike the concept of boundary, which I see as essentially privative, the concept of threshold is always also copulative. One could argue that the concept of the border is an incomplete concept of the threshold, because the border as a phenomenon always establishes a relation to what is beyond the border. But perhaps it is rather a question of emphasis, where the threshold concept more clearly underscores that it is precisely a phenomenon where what the threshold unites or separates belongs to the experience of the threshold as such? The border also unites by virtue of the fact that it separates. The border between two states can be perceived as a difference established between them. There is a border control, the function of which is to prevent unauthorised traffic between these two states. But the border control is actually part of a more extensive borderland that bears the characteristics of a threshold, since it is neither here nor there, but is an intermediate state that enables the passage between states and thus puts them in contact with each other. The border guard who checks my passport belongs to this intermediate state because he has already put me in contact with the nation I want to enter, even when I am denied entry. To think or draw a line is in a sense to have already crossed it. Thus, the border becomes a kind of threshold.

A threshold is not primarily a thing but a function or a symbol. The insignificant threshold slab that separates my office from the corridor outside would not really deserve to be called a threshold when it was lying by itself on the shelf of the DIY store. Possibly it could be called a potential threshold. Only when it is in place and can really distinguish between an inside and an outside in the form of an office and a corridor can it be called a threshold in a more definitive sense. It needs a specific inside and a specific outside to both unite and separate; its character cannot be separated from its function. Thresholds are not just boundaries between inside and outside taken on their own. They cannot be thought of as thresholds without also thinking about what they are thresholds between. Both conceptually and experientially, a threshold is inevitably accompanied by that which it unites and separates, abstractly or concretely.

The Ambiguous Thresholds

The elusiveness of the elusive thresholds has to do with their dynamic character. A threshold can be a physical thing, a "demarcation between rooms", but it is also a symbol of an existential passage or turning point. It is precisely its character as an intermediate or transitional state that gives it ambiguity as a phenomenon. We have seen how Homer, according to Porphyry, considered the gates of the cave of the Naiads to be sacred. From the point of view of human experience, this is surely related to the fact that, in these transitions, a human being is becoming someone other than what he or she was before, but has not yet become. A human being is on the threshold of an existential movement in which the foundations of his or her identity are being renegotiated. This unstable, elusive and precarious state has always been surrounded by various rituals because here we as human beings are in a vulnerable or dangerous situation. Being in the intermediate state of the threshold inevitably implies a loss of control, in which I am no longer an agent capable of controlling myself, my environment or the outcome of my passage. Instead, I relate here to the transition as the swimmer to the water; carried by the waves through which the swimming strokes propel me forward.

The very openness of this vulnerable state where we have to rely on circumstances contributes to the relative silence that surrounds thresholds. We generally do not want to draw attention to our loss of control. Moreover, modern society privileges our need for control over the realisation of dependence.[21] The concept of the symbol becomes important if we still want to make the existential meaning of thresholds visible. As much as the more stable places, the unstable thresholds structure our relationship with the spaces we inhabit. Thresholds are not only images of our existential passages but also, at least occasionally, their conditions of possibility and driving factors. It is probably the ambivalence of the threshold experience that allows thresholds to be perceived as uncanny, inadmissible or existentially crucial to cross. On the threshold, we are no longer the same and our claims to self-identity and self-control are questioned. If we are not quite ourselves on the threshold, we are not quite someone else either.

Perhaps one could say that thresholds are symbols of the changeability of human existence. This does not mean that every time I cross a threshold I get a new life. Not every change has to be of an existentially decisive nature. The anthropologist and sociologist Erling Goffman has made a distinction between how we appear in

21 Cf. my discussion on the buffered self and the porous self in Sigurdson 2021, pp. 243–256.

different ways *on stage* and *behind the scenes* – that is, in public and private contexts – and the switch between them is unforced and more or less unreflected, but nevertheless a passage.[22] Similarly, the anthropologist Marc Augé has described how the Parisian descends into a metro station as a private citizen and emerges from another as a worker or civil servant.[23] In both cases, it is a change of role directly or indirectly linked to and described by means of physical space. While these role changes are hardly existentially irrelevant, we can assume that in most cases they are not existentially decisive or transformative either. What is more constant about thresholds as a symbol of change is therefore precisely the passage as such, not the intensity of the change, which can vary from almost non-existent to being a passage to a new life.

In this introduction to thresholds, I want to try to understand the relationship between physically or geographically concrete thresholds and their symbolic meaning as an intermediate state or a turning point and how they structure the relation to the spaces we inhabit in both a concrete and existential sense. In the next chapter I will therefore turn to the question of existentially significant passages with the help of an anthropological discussion on passages and liminality. What is meant by a transition or a passage and how common are they in our time? According to anthropology, the liminal state – that is, literally the threshold state – is both spatially and symbolically structured. Therefore, it sheds light on passages in general and the experiences of change we make in them. Next, I continue the discussion on passages by discussing revolutions from the perspective of the history of ideas and political philosophy. Revolution can mean many things and take place in many different spheres – here the breakthrough of architectural functionalism in Sweden will serve as an example – but common to them all is a fundamental upheaval of the state of things in a certain area, which presupposes a threshold that somehow mediates between a before and an after. I then turn to a phenomenological investigation of our

22 Goffman 1956, pp. 66–86.
23 Augé 2002.

human experiences of thresholds, especially in conversions, therapies and limit situations, and what characterises these experiences. Based on these reflections in the mode of anthropology, history of ideas, political philosophy, phenomenology, and psychoanalysis, I then turn to the task of examining concrete, physical thresholds in space – the doors and windows of the house, buildings and streets in the city, borders between neighbourhoods – and their existential significance for our relationship to space. Here, finally, we will find ourselves on the threshold.

A recurring theme in all the chapters is the relationship between the physical and symbolic aspects of the threshold. As guides in threshold knowledge, we meet above all the French anthropologist Arnold van Gennep and the German writer, historian and philosopher Walter Benjamin. The importance of good guides to crossing the threshold should not be underestimated, as we will soon see.

In November 2019, while travelling from Gothenburg to Rome via Frankfurt Airport, I read Tim Ingold's *Lines*. For me, this journey, which I have made quite a few times, has taken on a personal meaning related to my work for several reasons: a pause, a semicolon between one place and another, which allows for a restart in writing. In the work of the anthropologist Ingold, however, I encountered a distinction that became a commentary on my own travelling. He distinguishes between two different forms of travelling: wayfaring and being transported.[1] Wayfaring presupposes an intimate connection between the movement of the journey and the perception of the journey; the journey is a passage, where the perception of the places I pass is part of the journey itself. Transport breaks this connection in its focus on the destination. Here I become a passenger who is passively transported and for whom the experience of the intermediate stations is indifferent. The journey itself becomes instrumental. It is about getting there.

I immediately realised that my journey to Rome was one of transport rather than wayfaring. Travelling in a more genuine sense, for example by train, would hardly have been possible, given all the work I had to do in Gothenburg, both before and after my trip. But Ingold also made me reflect on the difference that wayfaring and transport signify in relation to place. Transport, in its disregard for the significance of the concrete places to be passed, is *utopian* in character, that is, placeless. Wayfaring, on the other hand, is *topical*; it is concretely tied to a place rather than abstractly, because

1 Ingold 2016, pp. 78–85.

it presupposes that the relation to the places passed through is of importance for the journey itself. In this sense, the airport itself is utopian: it is built for transport and therefore has the character of a generic non-space. It is not a place I am invited to participate in, but a station I must pass through with the least possible friction.

Perhaps the airport as a stopover location should still be understood as a kind of threshold? Its position between departure point and destination suggests something like that. But unlike the thresholds I have discussed so far in the previous chapter, it works best when it affects the traveller as little as possible. For practical reasons, it is often most convenient to try to avoid stopovers and Frankfurt Airport hardly carries any existential significance, at least not for me. In Ingold's terms, origin and destination are two points between which the stopover airport serves only as an interface.[2] As such, it is not supposed to have any existential density of its own, which is another way of saying that it does not aspire to be a threshold in the true sense. Nothing of importance happens here. An airport is not built to be a transition or an intermediate state where I become, in some sense, another: "the place of transition to a new life".[3] Instead, the airport is meant to be a hypermodern non-place, a space without any identity of its own, lacking any semblance of its own "here" and therefore not incorporated into any meaning-making relations.[4] My stopover on this journey happened to be in Frankfurt, but it could just as easily have been in Copenhagen, Brussels, Amsterdam or Munich. It could have been anywhere, or rather nowhere at all. If something decisive had happened, it would probably have had no real relation to the airport as such.

According to Ingold, travelling in the form of wayfaring constitutes a fundamental way of inhabiting the world. For the passenger who is merely transported, the world is merely a blank surface filled with punctual destinations without any significant transitions in the form of thresholds. In reality, of course, there is no pure transport

2 Ingold 2016, p. 77.
3 Bollnow 2011, p. 150.
4 See Augé 1997, pp. 75–115.

without some element of wayfaring, not even for the air traveller: "We cannot get from location to location by leap-frogging the world, nor can the traveller ever be quite the same on arrival at a place as when he set out."[5] Travelling by plane is also a journey through time and space, and although the airport does not leave much room for the experience of being on my way somewhere else in a deeper sense, it does not prevent it from becoming for me an experience of entering a different way of being than before. Once at my destination, it becomes the gateway to an experience of no longer being "on the way" but instead "there".

For us humans, travelling can hardly be reduced to mere transport (and I am talking here about voluntary travel, not escape or deportation). Even in modern society, which, according to Ingold, is characterised precisely by a particular inclination towards transport and a generally utopian (in the sense of placeless) attitude, in which place is not supposed to play a decisive role, we are wayfaring in some minimal sense, because the passenger cannot leave his body and the space it inhabits behind.[6] Rather, the ideal of frictionless travel – travel in which the places we pass through are irrelevant to travel itself – means that the spaces through which we travel are actually planned and built in a way that favours transport over wayfaring. The experience of wayfaring is diluted and subjective because it is not supported by these spaces, but it does not disappear. The airport or the bus station or the train station – if planned and built according to this ideal – is not conceived as a threshold or a gateway but as a pure interface.[7] The ideal is that I remain in all respects the same at the destination as I was at the starting point. With intermediate stations as frictionless as possible, it simply becomes more difficult for the journey as such to constitute a unique experience. The loss of location and the loss of experience go hand in hand.

---

5 Ingold 2016, pp. 105 f.

6 See also what Henri Lefebvre has to say about abstract and transparent space in Lefebvre 1991.

7 On the railway station as a gateway, see Sigurdson 2024c, pp. 127–136.

In a work aptly called, in German, *Das Passagen-Werk – The Arcade Project* in English –, Walter Benjamin has argued that in our modern age "[w]e have grown very poor in threshold experiences".[8] The reason for this is that genuine passages are "becoming ever more unrecognizable and impossible to experience". What Benjamin is pointing to here is just such a dilution of experience that is the consequence of places playing less and less of a role in our passages. Rites are a way of anchoring ourselves in time and space, but they are also a pause, a semicolon, which separate a before and an after and therefore denote a transition or an intermediate state that is not just a flat contact surface but a "zone". For Benjamin, the threshold has both a symbolic and a concrete meaning:

> Transformation, passage, wave action are in the word *schwellen*, swell, and etymology ought not to overlook these senses. On the other hand, it is necessary to keep in mind the immediate tectonic and ceremonial context which have brought the word to its current meaning.

In this poverty of threshold experiences, Benjamin highlights falling asleep and waking up as examples that are still valid for his time – and perhaps for our own. Should we understand the choice of these as an example of a modern distinction between place and meaning, since falling asleep in our time is rarely accompanied by any ritual other than in a figurative sense, and waking up is often not accompanied at all? Or is it a realisation that threshold experiences are in some sense inevitable, even when they have become "ever more unrecognizable and impossible to experience"? Sleep is an experience where we have to give up our own initiative and control, a bit like a ritual.[9]

8 Benjamin 2002, p. 494. All quotes in this paragraph are taken from this page.

9 See Crary 2013, p. 125.

Rites of Passage

"*Rites de passage*" – this is exactly how Benjamin begins, in French, his paragraph on threshold experience. Benjamin's *Arcades Project* was written between 1927 and 1940 and deals with the arcades that existed in Paris in the 19th century. It has an intricate structure of fragmentary analyses, quotations and reflections through which Benjamin attempts to describe the city's arcades as a symbol of emerging modernity. It also includes a page of reflections on threshold experiences in relation to passage, and thus begins with what should be a reference to the French anthropologist Arnold van Gennep's epoch-making 1909 work, *Les rites de passage*, although Benjamin does not mention van Gennep by name. van Gennep was the one who established the concept of "rites of passage" and in Benjamin's explanation of rites of passage, they are "the ceremonies that attach to death and birth, to marriage, puberty, and so forth", a selection of the rites of passage van Gennep discusses in his book.[10] It is as transitions that these ceremonies "are becoming ever more unrecognisable and impossible to experience". In Benjamin's work, "ceremony" seems to be used synonymously with "rite".

The allusion to van Gennep is interesting, since in the first half of the 20th century this anthropologist was not yet the obvious reference in the field. van Gennep (1873–1957) was a French anthropologist, ethnologist and folklorist with eminent language skills and an extensive list of publications. He was instrumental in establishing anthropology and sociology as sciences. Despite his, in retrospect, undeniable competence, he was never given an academic position in his native France because the opposition of who was, at the time, the dominant sociologist, Émile Durkheim (1858–1917).

Why did Durkheim want to keep van Gennep out of the academy? One answer is that van Gennep stood for a competing idea of the social.[11] *Les rites de passage* is essentially a comparative analysis of concrete rites of passage, but it is more than that: it is an interpretation

10 Benjamin 2002, p. 494.
11 See Thomassen 2014, pp. 21–70.

of the significance of these rites for the meaningfulness of life. Rites of passage have not only a personal significance but a social one. Indeed, it could be argued that in reflecting on rites of passage and experiences of passage, van Gennep's book aimed to present a more general idea of the social, which is about how rites mark the transitions that occur always and everywhere in all societies. Rites of passage were thus not marginal ceremonies but central to van Gennep's understanding of human society.

van Gennep's account of rites as central to human society was based on extensive and systematic empirical studies of the patterns that occur in the rites. This meant that van Gennep took a different view from Durkheim on social change and criticised him for reducing any explanation of it to a function of society, a criticism that had also been levelled at Durkheim earlier by the French sociologist and social psychologist Gabriel Tarde (1843–1904). The weakness of Durkheim's idea is that society tended to be understood as an entity beyond the lives of individuals that could be used to explain everything.[12] But for van Gennep, rites functioned simultaneously on an individual and a collective level; they were both experiences for the individual and symbolised social transitions. Therefore, the rites were very much alive and changing; they were patterns for these transitions rather than abstract laws. In other words, the rites of passage structured the relations between people in a society in a way that was at once concretely material and symbolic, both "objective" and "subjective" if one takes a contrastive view of them. Among the differences between van Gennep and Durkheim is also van Gennep's emphasis on the importance of the liminal – the threshold – something that Durkheim did not pay much attention to. van Gennep's book on the rites of passage therefore came to challenge Durkheim by virtue of a different way of thinking about the social, in which individual and society are not played off against each other but are united by virtue of the rites as social interaction in the form of transition or, if you like, threshold experiences.

---

12 See Asplund 2022, pp. 319–390.

In 1960, van Gennep's *Les rites de passage* was translated into English and subsequently established itself as an anthropological classic. Influential British anthropologists such as Victor Turner (1920–1983) and Mary Douglas (1921–2007) are among those who were fundamentally influenced by van Gennep's idea of the rites of passage and the liminal state and thereby have come to influence the book's reception.[13] Through Turner in particular, the concept of liminality has also become a central part of the attempt to understand human society. His essay "Betwixt and Between" draws attention to the liminal not only as a transition but also as an intermediate state that is connected to and dependent on space. The term "liminal" itself comes from the Latin *limen*, which means border or threshold. For Turner, as later for Douglas, the threshold was a key to understanding society, although Turner's theory of the liminal came to extend van Gennep's notion of the importance of the intermediate state and threshold experiences. When Benjamin referred, at least indirectly, to van Gennep's work a number of decades earlier, he anticipated, at least in part, a realisation of the groundbreaking nature of van Gennep's concept that would be developed more widely much later.

According to van Gennep in *Les rites de passage*, rites of passage are essentially about the transition from one state to another, where the differences between these states are so great that an intermediate state is necessary for the passage to succeed. The first example van Gennep mentions is passing between the profane world and the sacred, but it soon becomes clear that it is not just a matter of a religious division of the world apart from everything else. Every minor and major change of cosmic or social world presupposes a similar transition: birth, adolescence, marriage, parenthood, change of class, occupational specialisation, retirement, death, but also lunar phases, seasonal changes and the New Year. In different parts of the world, the varying elements of such a transition may be given different significance or emphasis, but essentially the rites of passage can be divided into three types or subcategories that together describe a completed transition or a sequence that is

13 See Turner 1967, pp. 93–111; Turner 2017; Douglas 2001.

almost a rudimentary narrative with a beginning, middle and end, or a journey of departure, travel and arrival: rites of separation (*rites de séparation*), transition rites (*rites de marge*) and rites of incorporation (*rites d'agrégration*).[14] In this context, van Gennep also uses a terminology that, linguistically speaking, centres on the threshold itself by dividing the types into pre-liminal, liminal and post-liminal rites. Since these are passages in which the social status of the people involved changes decisively, they affect not only the individual but also society. As each change is associated with both danger and possibility, the function of rites of passage in relation to these changes, writes van Gennep, is to "reduce their harmful effects".[15] Although there is great variation and different cultures place different emphasis on one or more of these elements, the same pattern emerges across the world. In many of these passages, the intermediate state or liminal period is an independent element.

The transition that takes place in the rite of passage is linked to a physical transition. In the second chapter of *Les rites de passage*, entitled "Le passage matériel", van Gennep argues that those who undergo an existential passage also undergo a spatial transition. A "symbolic and spatial area of transition" is somehow present in all rites of passage from one state to another.[16] The book mentions a number of potential transitions of this kind, from gates and statues to zones between countries, territories, villages, towns, neighbourhoods, temples and houses. Finally, the "neutral zone shrinks progressively till it ceases to exist except as a simple stone, a beam, or a threshold (except for the pronaos, the narthex, the vestibule, etc.)".[17] Here, then, the threshold appears rather as a remnant of a more clearly physically marked transition, where the gate surrounded by taboo prohibitions eventually turns into an ordinary door into a house. Nevertheless, the door and its components retain a certain sacredness, from the threshold to the lintel and architrave, and therefore also certain rites.

14 van Gennep 1981, p. 20; van Gennep 1960, p. 11.
15 van Gennep 1981, p. 22; van Gennep 1960, p. 13.
16 van Gennep 1981, p. 28; van Gennep 1960, p. 18.
17 van Gennep 1981, p. 29; van Gennep 1960, p. 19.

The door is "the boundary between the foreign and domestic worlds in the case of an ordinary dwelling, between the profane and sacred worlds in the case of a temple".[18] Crossing the threshold means joining "a new world" in both the former and the latter cases, which is why so many ceremonies are linked to the threshold as such in, for example, weddings, adoptions, ordinations or funerals.

The rites performed on the threshold are "transition rites", in van Gennep's terminology. They are therefore neither rites that separate a human being from his or her previous stage nor rites that unite him or her with a future stage, but rites on the margin between the old world and the new, rites "executed during the transitional stage" or liminal rites.[19] For van Gennep, the terms "marge" and "seuil", i.e. "margin" and "threshold", seem to be more or less synonymous. The "margin" is not something marginal, that is, something that exists alongside what is more central, but precisely a margin between two states and, in other words, a threshold. In the concluding paragraph of the second chapter, van Gennep writes about "rites pertaining to the threshold" to emphasise that the threshold is only part of the door.[20] These rites are not about the threshold as a physical thing, although it will take on a symbolic meaning in the rites of passage, but precisely about passages: entering, waiting and leaving. Sometimes, in the case of the threshold of a house, these rites apply only to the main door, while other entrances may be less symbolically charged in terms of the transition between a foreign and a familiar world. In other cultures, each threshold carries a sacred charge.

Although it plays an essential role in the transition between the sacred and the profane, van Gennep's example makes it clear that it is not only about the thresholds associated with what is called religion in our time. All the foreign and familiar thresholds of the everyday world also seem to be loaded with symbolic meaning. Perhaps van Gennep is hinting at a difference between our culture and "civilisations other than our own", where thieves in the latter tend to

18 van Gennep 1981, p. 29; van Gennep 1960, p. 20.
19 van Gennep 1981, p. 30; van Gennep 1960, p. 21.
20 van Gennep 1981, p. 34; van Gennep 1960, p. 25.

avoid the main entrance because of its deeper symbolic charge, and dead bodies are carried away via the back door and windows for the same reason.[21] Is this a gentle hint that passageways play less of a role in our modern culture, as Benjamin, for example, claimed?

A society can be likened to a house, according to van Gennep, with many different rooms and corridors. The more similar a society is to our own culture, the thinner the walls separating these spaces and the wider and more open are the "doors of communication" between them.[22] In more traditional societies, rites of passage are more common. van Gennep thus seems to argue, like Benjamin, that rites of passage "have become increasingly unrecognisable and unperceived". Nevertheless, in his conclusion, van Gennep reiterates that the social organisation of human societies is based on distinctions between social groups that require some kind of transition or physical passage to be bridged:

> For groups, as well as for individuals, life itself means to separate and to be reunited, to change form and condition, to die and to be reborn. It is to act and cease, to wait and rest, and then to begin acting again, but in a different way. And there are always new thresholds to cross.[23]

That there are similarities between the rites performed has been shown by many.[24] What van Gennep believes he has clarified is that such passages follow a specific pattern, with the transition or threshold state itself constituting a more ambiguous intermediate state, separate from both separation and initiation. In the intermediate state, the subject is ambivalent and cannot be defined by either the preceding or the following state because it has no known coordinates to adhere to. This intermediate state is in turn associated with a "passage matériel" – a physically concrete passage.[25] The physical passage is thus not only an image of a more existential passage, but

---

21 van Gennep 1981, p. 34; van Gennep 1960, p. 25.
22 van Gennep 1981, p. 35; van Gennep 1960, p. 26.
23 van Gennep 1981, p. 192; van Gennep 1960, p. 189.
24 Cf. Trumbull 1896, which van Gennep both builds on and criticises.
25 van Gennep 1981, p. 194; van Gennep 1960, p. 192.

a symbol of it, in the sense I use the concept of the symbol here. Crossing the threshold to a new world and becoming another is associated with also crossing a physical threshold in some form. However thin this threshold may be in our time, according to van Gennep, it is not non-existent. Thresholds remain essential to the social organisation of the social house into different spaces in both an existential and a physical sense. van Gennep does not really study the intermediate states of our own culture, although he believes that the recognition of these is one of the essential points of talking about passage rites. However, his followers would do so.

From the Liminal to the Liminoid and Back Again

Addressing van Gennep's idea of liminality as part of a passage in a 1964 lecture, Victor Turner argues that although rites of passage are expressed in a particular way in traditional, more cyclical societies, they exist in all societies.[26] However, in this context – written just after discovering van Gennep's book – he applies the concept of liminality to the initiation rites of the Ndembu people of Zambia, rather than his own culture. Turner, like Mary Douglas at the same time, emphasises the ambiguity of the liminal state. It is an "ontological transformation", not just a transport from one position to another where nothing essential changes.[27]

In Douglas, who like Turner explicitly draws on van Gennep, the relevance of the liminal category to modern society is evident and the ambivalence of the state is striking. The intermediate state is a state that is associated with both power and danger: "Danger lies in transitional states, simply because transition is neither one state nor the next, it is undefinable."[28] During the liminal period, what Turner calls an "ontological transformation" takes place, rendering the person both individually and socially ambiguous. Being able to get through such a transformation of one's entire existence is not

26 Turner 1967, p. 93.
27 Turner 1967, p. 102.
28 Douglas 2001, p. 97.

something that can be assured on its own. The person in the passage is a danger, both to himself and to others. The function of the rites is to manage this danger, to ensure that the transition goes well and is completed.

However, in a society such as the modern one, with its relative absence of rituals, this absence means that those who are in a threshold state also risk remaining in it. Remaining in the threshold state means being exposed to a perpetual disorder in which vulnerability to manipulation by malevolent forces increases and in which it becomes impossible to get to know one's society, its social relations and ultimately oneself in a genuine way, because fixed contours are lacking.

Turner was also quick to apply van Gennep's three-part scheme to his own society in his 1969 book *The Ritual Process: Structure and Anti-Structure*. Unlike Douglas, however, Turner believes that the liminal state in the modern world has become a more or less permanent condition.[29] In the mendicant and monastic orders of world religions, but also in the court jesters and the chiliastic movements, there are early versions of such an institutionalisation of the intermediate state. Often, those in this state have come to be regarded as sacred or endowed with magical powers and therefore as somewhat dangerous. At the same time, Turner argues that these structurally marginal figures represent an "open" morality by virtue of their ambivalent existence. They represent a social formation that Turner calls *communitas* or an "open society" that potentially embraces all people equally, at least in contrast to the prevailing and "closed" society with its fixed structure and hierarchical organisation. Here people meet each other outside their social roles, face to face.

Communitas and structure are, for Turner, dimensions of all societies, including the modern one. For his part, Turner was particularly interested in the hippie movement in North America in the 1960s, which he believed embodied many of the values that characterised *communitas* in his own time: spontaneity

29 Turner 2017, p. 107.

and immediacy.[30] The modern pilgrimage also bears features of *communitas* in that it establishes a community outside the framework of established society.[31] *Communitas* and social structure are both dimensions of society, but their relationship to each other is one of dialectical contrast rather than pure antagonism. The liminal and marginal was thus an integral part of modern society, and without the creative ambiguity that this condition represented, society would never be renewed. Such a condition was hardly ever a stable condition, but rather a form of purification and renewal of the social structures.[32] Without such liminal intervals, no structure could exist either. Social life is a dialectical activity, a movement from structure to *communitas* and back to structure.

In this way, Turner extends van Gennep's scheme of passages to a full-scale theory of society. If van Gennep concentrated mainly on smaller societies, Turner came to explicitly apply his theory to larger and more modern societies. In doing so, he saw himself as applying van Gennep's insights into the transitions in societies that are associated with seasons and therefore concern the whole of society and not just, as van Gennep himself did, transitions related to personal life crises.[33] There is, however, according to Turner 1974, a crucial difference between traditional societies and modern societies after the industrial revolution in terms of this intermediate state. If traditional societies are characterised by a relative stability in which the social system is cyclical and repetitive, this means that the liminal state is also inserted into a cyclical movement in which this ambiguous state is reintegrated into the more stable structure after a period of time. If the liminal state is associated with "play", this is a play whose scope and meaning is nevertheless determined by the "work" represented by the social structure that envelops the whole process. The "play" of the liminal state here, unlike in society after the post-industrial revolution with its emphasis on innovation and

30 Turner 2017, p. 113, cf. pp. 132, 138 f.
31 Turner and Turner 1978, pp. 37–39.
32 Turner 2017, p. 185.
33 See Turner 1974, p. 56.

change, cannot unbalance the social structure. In modern society, on the other hand, "play" will become an activity alongside "work", a relatively "independent domain of creative activity".[34] By extension, such a domain may exhibit other models of living that can influence and change the way of living in the established society in a decisive and irreversible way. The difference between traditional and modern society seems, for Turner, to be about the difference between a cyclical and a progressive-linear society.

To distinguish the intermediate states of traditional and modern society, Turner introduces a distinction between the "liminal" and the "liminoid" (i.e. that which resembles the liminal), the latter being typical of the modern. In the liminal state of traditional societies, it is in principle impossible to separate "work" from "play", whereas modern society has produced a crucial distinction between "work" and "leisure", where leisure is by definition a time without work. If leisure is defined as time without work, it changes the meaning of all kinds of play, including ritual. It becomes a part of a person's life that is autonomous and voluntary. The difference between the liminoid and liminal states is therefore characterised, among other things, by the difference between the optional and the obligatory: "One is all play and choice, an entertainment, the other is a matter of deep seriousness, even dread, it is demanding, compulsory".[35]

The liminal state, where all the accepted categories are questioned, has in principle secreted the liminoid that has been waiting until the social structure allows it to reverse the relationship between structure and antistructure. When the ambiguity of the intermediate state makes the outcome precarious, then there is also the germ of a radical social and cultural change and not just a return to an already established structure. Social revolutions are liminal phases for which the threshold states of the transitory states, as well as *communitas* in various forms, are harbingers or premonitions. But the crucial difference between the liminal and the liminoid remains: in the former, the intermediate state is part of a passage but in the latter

34 Turner 1974, p. 65.
35 Turner 1974, p. 74.

it coexists with the social structure, in the former it is obligatory but in the latter optional, in the former collective and in the latter mostly individual. Unlike van Gennep's understanding of the liminal phase, the liminoid state is not really a transition or passage, although it may affect society through its other ways of life. Rather, it is a dialectically present possibility. Therefore, the liminoid state is not necessarily associated with any rites by means of which it becomes possible to orientate oneself in it. On the contrary, the rite is transformed into a liminoid game. The rite, in the liminoid state, turns into a pretended staging that belongs to the sphere of choice rather than necessity, more a representative and creative expression of myself than an experience that I am involved in but do not possess myself.[36] Even as traditional rites persist in modern societies, their meaning changes from social event to individual act.

With Turner's distinction between the liminal and the liminoid state, van Gennep's understanding of liminality has expanded but also changed. For van Gennep, the liminal state in the passage is undoubtedly an intermediate state, while Turner's liminoid state takes place more at the margin. van Gennep repeatedly uses the term *marge* and, as I pointed out above, more or less as a synonym for *seuil*. But although the margin and the threshold are both concepts signifying a border or a limit, they occupy completely different positions. An intermediate state, such as the one van Gennep speaks of, is, by virtue of being in between something, also in the centre of something.[37] Whether or not the status of the marginal in modern society has been reassessed from threat to creative possibility, the liminoid, in Turner's version of it, takes on more the character of an ever-present possibility that stands in a dialectical relation to an established social structure and is thus "beside" rather than "between". Something essential has changed in relation to van Gennep's understanding of the passage.

As I have already mentioned, liminoid experiences in modern society are experiences that are voluntary and self-initiated. They

36 See my discussion on this in Sigurdson 2025a.
37 See Thomassen 2014, pp. 7 f.

do not constitute a station in a ritual passage and do not offer an experience that seriously changes the very foundations of existence. In the liminal state, on the other hand, its threshold character means that the passage is a transformation, a movement across different guises or forms, which has no stability of its own but exists by virtue of what the movement both unites and separates. The liminoid is similar to the liminal, as the literal meaning of the term suggests, but more as a play with transformative experiences than by offering experiences that are transformative in the true sense. Jumping from a great height with only a strong rubber band attached to your feet – bungee jumping – may be, as the anthropologist Bjørn Thomassen points out, a game with death and as such a liminoid experience.[38] But unlike the liminal experiences that the passage initiates and accompanies, the rubber-banded leap does not involve a symbolic death, where the subject after the experience becomes in some fundamental sense different in the eyes of society as well as in his own. Instead, the recoil of the rubber band will immediately pull the jumping person back to normal. Bungy jumping is a liminoid kick, not a liminal experience.

A liminal passage is, as Douglas and others point out, fraught with danger. But here too the liminal is different from the liminoid. Of course, bungee jumping can be dangerous, even if the safety regulations are (one hopes) rigorous. But the danger associated with it is primarily physical, not symbolic, death. Crossing the threshold is instead about danger on a symbolic level. To remain in the intermediate state if the crossing fails is to be in limbo. The intermediate state involves a transformation of the very foundations of existence. Such a transformation implies a symbolic death of the previous identity, for example in adulthood where we are no longer children, as a prerequisite for a transformation and new symbolisation of life.[39] As long as we remain in the transitional state of adolescence and have not yet reached a postliminal state and a new identity by becoming adults, our existence will be ambiguous and fluid. Those

38 Thomassen 2014, pp. 167–190.
39 See Szakolczai 2015, pp. 17–19.

who do not pass through a passage well risk remaining indefinable, as if forced to live in a permanent adolescence. In a more societal sense, it corresponds to a state of permanent revolution. Although Turner also recognises that the transition is fraught with danger, he tends to emphasise the liminoid state as something positive by associating it with creativity and play, with *communitas* as an egalitarian community and social emancipation and renewal. For Turner, the liminoid seems to be not only a descriptive category but also a normative ideal.[40]

Such idealisation risks underestimating the perilous aspects of the liminal state by downplaying both its transitory nature and the rituals whose function was precisely to protect against danger by guiding. But even when the perilous aspects of the liminal state are mentioned by Turner, it is the creative possibilities that play the main role.[41] Turner has a point, of course, in emphasising how the liminal state potentially carries emancipatory, creative and renewing features. But by highlighting the ambiguous state as such as an ideal, he either ignores its potentially destructive aspects in the form of fragmentation, existential and political disorientation as well as incurable restlessness, or reduces it to mere entertainment that does not entail any real change. A state of flux is not only an opportunity for constant creative reorientation but can also give rise to confusion, anxiety and rootlessness. The flip side of Turner's liminoid state is what sociologist Zygmunt Bauman calls a "liquid modernity", that is, a modernity that is no longer a break from past traditions towards a new, modern form, but a constant break, also from the limits previously set by the constraints of space, characterised above all by anomie.[42] In liquid modernity, we travel through life as tourists, without any permanent ties to places, people or identities, "being unable to stop and even less able to stand still". Therefore, we do not

40 Thomassen 2014, p. 10.
41 See Thomassen 2014, pp. 83–85, 185–190.
42 Bauman 2000, pp. 28 f.

move toward any final destination, but remain engaged in a constant game of musical chairs where no place is permanent.[43]

It is certainly to Turner's credit to have shown how the liminal state affects modern society and not only traditional societies. At the same time, van Gennep's insights into the transformative character of the passage rites risk being lost, as does the realisation of how they structure society as such and not just the part of society that belongs to the cultural sphere. Even in modern societies, we may assume, transitions take place in the form of passages from child to adult but also, for example, at birth and death. And even collective events, for example in the form of major sporting events or religious festivals, often assume a ritual character. If passages involve a fundamental existential change on an individual as well as a social level, then Turner curiously leaves aside the observation that van Gennep makes throughout his *Les rites de passage*: the liminal as an intermediate state and the rites specific to this state as a companion in this existentially precarious situation.

van Gennep's book is not about presenting a theory of rites of passage but about uncovering a common pattern that runs through an impressively large number of different rites, in which the intermediate state plays a crucial role. It is through these that individuals and societies are moulded, and if they do not take place or if they fail, then this also affects both the individual and the society. It is crucial to the understanding of liminal experiences that they do not fall entirely within the subject's own power. Rather, they are about experiencing something that is not under one's own volitional control. Turner was, of course, not unaware of this. In his 1964 lecture on the liminal, he emphasises precisely this as a state "betwixt and between", an intermediate state, not merely an object of our choice and control, but as such something to which we stand in a passive relation.

However, the obvious risk of seeing the liminoid, as Turner does, primarily as an outflow from the liminal, and even as its replacement in modernity, is to understand modernity as forever trapped in what anthropologist and sociologist Arpad Szakolczai calls a "permanent

43 Bauman 2000, pp. 28, 33 f.

liminality" – a term that captures the present perhaps even better than "liquid modernity".[44] A permanent liminality is a passage without an end, and therefore no passage at all, since there is no threshold between a before and an after and no real transformation ever takes place. Living in a permanent liminality can be likened to the living dead. Or, to choose a less drastic analogy, permanent liminality is more like transport than a journey on the road of life. Temporary liminal situations are transformed into permanent marginal identities. Such a literal marginalisation of liminality can be understood precisely as a defence against the liminal and its potentially transformative character. To put it more precisely: the discontinuities that liminality necessarily implies become, for historical and political reasons, problematic, so in order to ensure continuity and conformity at the centre, they are relegated to the margins. In order to pacify the transformative nature of liminal passages, they are transformed into a non-committal game alongside the seriousness of economics and politics. They become existentially cool and optional ceremonies, instead of existentially hot and invocative rites. As Szakolczai puts it: liminality is marginalised and the marginal is liminalised when modernity begins to understand a human being in terms of their "true self" instead of a narrative of "the sequential rhythm of liminal processes".[45] By making liminal situations permanent, one secures oneself against the disorder and potential danger that their transformative character actually implies.

Again, it was van Gennep's central insight that passages, in order to be precisely passages where something decisive happens, must be tripartite processes, not binary oppositions between order and disorder or continuity and discontinuity. If the liminal is relegated to a marginal position alongside the centre, it becomes a state of disorder alongside a central order, but, as Szakolczai notes: "If everything is continuously changing, then things always remain the same."[46] The question therefore becomes whether the liminal, in van

44 Szakolczai 2000, pp. 207–217; Szakolczai 2017.
45 Szakolczai 2000, p. 188, cf. p. 184.
46 Szakolczai 2000, p. 217.

Gennep's sense, still exists in our time. Are there any significant and distinctive thresholds left?

Liminal Walks

I think once again about my trip to Rome. What if I had walked to Rome instead of flown? According to an Internet search engine, it is just over two thousand kilometres and would take four hundred and twenty-seven hours, or just under eighteen days. I don't think the search engine included any breaks, other than the ferries between Helsingborg and Helsingör and between Gedser and Rostock. So let's say, for the sake of calculation, that I could walk for eight hours a day and that I need one day's rest per week. Then the whole journey would take just over sixty days. Thirty-three kilometres a day, six days a week, sounds pretty reasonable. But it doesn't work. I don't have that much holiday in a year and I'm so busy at work that I couldn't be away for that long. Moreover, I haven't even factored in time for work in Rome and travelling home, nor time for getting lost nor for other possible delays along the way.

But still, what if I were to walk to Rome? I would, according to the proposed route, pass through Denmark, Germany and Austria. In Germany, I would walk through Halle in the state of Saxony-Anhalt, with its famous university, the city that was a centre of both Pietism and Enlightenment in the late 18th and 19th centuries. I would also pass through Munich in the state of Bavaria, captured in 1632 by Gustavus Adolphus, the king who founded Gothenburg in 1621. In Austria I would pass through Innsbruck, where I had never been, and in Italy, Trento, where the Tridentine Council took place in the 16th century. Then I would see beautiful Verona, with its Roman amphitheatre, Bologna, where I could eat tortiglioni in brodo and wander under the arcades, and finally Renaissance Florence before reaching Rome.

That doesn't sound so bad when I think about it.

Imagine how much I would experience along the way. The route even allows me to walk for a while on the Via Flaminia, the road travelled by the ancient pilgrims to Rome to eventually enter the Holy City via the Porta del Popolo. This is where the city of Rome opens up to the visitor, ever since Pope Sixtus V, towards the end of the 16th century, laid out his new city plan with three streets radiating out from Piazza del Popolo with Via del Corso in the centre. I wouldn't get that far, though, as I would probably make sure to live on Via Omero in Villa Borghese, that is, "outside the walls". But in a way I would be repeating a medieval pilgrimage *ad limina apostolorum*, "to the threshold of the apostles" (and there is the threshold again), the purpose of which was to visit the tombs of Peter and Paul. My main purpose would be to get a week or two of peace and quiet, but one need not exclude the other.

There are some practical problems with such a journey: the first thing that comes to mind, as I said, is that I would have to take too much time off work. But wouldn't it be possible to claim that the journey itself would be a kind of labour, as Nietzsche suggested?[47] Not only in the sense that travelling so far can be exhausting, but above all in the experiences – encounters with people, cities, history, challenges – that I would have. "Exploring the world is one of the best ways of exploring the mind, and walking travels both terrains," claims author Rebecca Solnit.[48] This sounds relevant to my own work. Let's say I'm about to write a chapter on passages in a book about thresholds. Wouldn't the experiences I undoubtedly would make during such a journey contribute much more to the intentional depth of what I am writing about than the transport of air travel? Wouldn't it be a kind of passage I would experience in that case? In contemporary literature on walking, there is a recurring idea that walking, including the hardships it entails, does something to us so that we are not the same when we arrive as when we started.[49] If travelling by plane is transport, walking is wayfaring. This is also one of the points of the pilgrimage, that something happens to us

47 Gros 2009, pp. 21–45.
48 Solnit 2000, p. 14.
49 See Solnit 2000, passim; Beaumont 2016; Gros 2009; Horvath and Szakolczai 2018.

while we are travelling.[50] Although I do not suggest that we are all on our way to the doorstep of the apostles in Rome, there may be something of pilgrimage in other journeys of a similar nature, where one's own hardships and physical locations interact on the way to the destination of the journey, which may be concrete, but also more symbolic.

Even being adrift can, for some, act as a passage. In August 2017, Christian Lewis, a former paratrooper with financial problems and mental health issues, set off from Swansea, Wales, on a six-year walk along the UK coastline, a journey of almost eighteen thousand kilometres.[51] After his military service, he had struggled to reintegrate into society. When his sixteen-year-old daughter moved out, he realised that his life had to change radically. With rudimentary equipment and no real financial means, he set off on a journey. Lewis's story is a contemporary tale of how someone puts himself through an ordeal in order to restore his relationship with himself, with his daughter and with his fellow human beings, and thus also with a community. Having a leaking tent, not enough batteries and very little to eat made him dependent on the kindness of strangers, and in this way he also managed to restore his trust in other people.

Lewis's story is not a story about being adrift in some entirely aimless sense, although the point is not to return to Swansea itself, which it seems he certainly plans to do. Rather, the point is that the wandering is a passage that follows van Gennep's three-part pattern, including its final phase, the phase of aggregation or return. There is also here, as Solnit puts it, as with the pilgrimage, a "symbiosis between journey and arrival".[52] Of course, Lewis is no longer the same person he was when he left. This is the very point of the ordeal. In Lewis's story, it becomes clear that there is a before, an in-

50 Cf. Turner and Turner 1978, pp. 1–17, 34–39, 231–237, 253 f. Again, I think Turner's distinction between the liminal and the liminoid is problematic. van Gennep uses pilgrimages as an example of passage in 1981, p. 187; van Gennep 1960, p. 184.

51 Lewis 2023; cf. Braidwood 2023.

52 Solnit 2000, p. 54.

between and an after, where Lewis becomes a different person. His walk is a kind of pilgrimage in that it has transformed both himself and his relationship with the world. Such stories of life-changing journeys are not uncommon. The walk in this sense is closely related to the rite of passage in that both embody a distinctly liminal state. Perhaps more than anything else, it embodies the transformation of the passage in a concrete and physical way. As Arpad Szakolczai and Agnes Horvath note in their book on walking: "Walking is a dangerous state; it is a liminal, in-between condition, implying a contact with the void."[53] Of course, such a characterisation does not cover every little walk we take. But it does suggest why walking can have a transformative function: we put ourselves at stake.

That Lewis's story fits so well into the pattern established by van Gennep suggests that passage is still functional today. It responds to a basic human experience that genuine change is indeed possible. Lewis's story is a story about the liminal, not about the marginal or liminoid. It is about passage as experience, as threshold experience. The threshold in his walk is certainly not as tangible as for those who cross the Porta del Popolo on their way to Rome, but nevertheless the relationship to the places he passes plays a crucial role in his transition. The existentially significant walk is not absent in our time. This fact suggests both that there may be other passages and passageways in our culture as well, and that modernity's relegation of liminality to the margins obscures rather than erases their presence. After all, thresholds are themselves an elusive phenomenon. Through passages we become aware of their function as intermediate states and transitions.

53 Horvath and Szakolczai 2018, p. 173, cf. pp. 34, 161–170, 173–185.

"All that is solid melts into air". This is one of the most famous phrases of Karl Marx and Friedrich Engels in *The Communist Manifesto* of 1848. And it continues: "all that is holy is profaned, and man is at last compelled to face with sober senses, his real conditions of life, and his relations with his kind."[1] This formulation captures the dynamic which, according to Marx and Engels, characterises the modern or "the bourgeoisie epoch", which is the preliminary culmination of a revolutionary upheaval.[2] *The Communist Manifesto* has undoubtedly come to offer one of the most influential inflections for understanding the dynamics of revolution in our time, although there are others. Marx and Engels describe revolution as a "reconstitution of society at large", a "forcible overthrow of all existing social conditions".[3] It arises out of the historical antagonisms between different social classes over access to the means of labour required to produce the objects by which our lives are sustained. Through historical development, as property passes from the hands of the princes to those of the bourgeois class and then to the working class, a state of affairs is eventually reached in which the historical antagonisms are resolved. Only through communism, where private property becomes the property of all, will these contradictions as such be cancelled. Only then will a society emerge "in which the free development of each is the condition for the free development of

---

1 Marx and Engels 2002, p. 244. For the circumstances surrounding the creation of the manifesto, see Liedman 2015, pp. 245–294.
2 Marx and Engels 2002, p. 244.
3 Marx and Engels 2002, pp. 219, 258.

all".[4] Historical development, according to Marx and Engels, is not an unbroken process leading straight ahead to the classless society, but a dialectical development that proceeds by detours of a more revolutionary nature. Historical development is not "evolutionary" but "revolutionary", to use a distinction that can certainly mean different things at different times.[5] History moves forward haltingly through violent upheavals in the political philosophy of Marx and Engels, but it moves forward nonetheless.

What I want to emphasise in Marx and Engels is not their philosophy of history as such, but the dynamics it exemplifies, which we also find in other concepts of revolution. Revolution is a "reconstitution", an "overthrow" and thus a break in the flow of history. Modern society, as a consequence of the bourgeois revolution, is a society in which "all that is solid melts into air". Marx and Engels, as we have seen, are careful to emphasise that revolution is a transformation of *the whole* of society, of "*all* existing social conditions", and not just a change affecting some parts of society but not others. In other words, revolution is a kind of passage, where the threshold between what came before and what comes after constitutes a particular state of transition. It is during the process of revolution itself that "all that is solid" has to "melt into air" in order to be different afterwards. Revolution is, to borrow a term from the last chapter, a liminal state. This applies not only to political and social revolutions, but also to economic, religious, technological and scientific revolutions. The industrial revolution at the end of the 18th century radically changed not only production but also living conditions, and thus also what it actually meant to be human.

The idea of a radical revolution raises the same questions as other liminal states: is there any continuity between the state before and after a revolution, or is it a complete new start? Is there an end to the revolution, when we can return to something solid and permanent? And who is actually the master of ceremonies, leading and overseeing the passage of the revolution itself, when the change of

4 Marx and Engels 2002, p. 244.
5 See Jonsson 2022, pp. 581–591.

transition makes us both malleable and vulnerable? Given that there is hardly any aspect of contemporary existence, neither existential nor material, that has not been touched by the last two hundred years of discussion on revolution, I remind you of the focus of my investigation here: the thresholds. Can the concept of revolution, not least the Marxist one, help us understand something about the liminal state and the threshold experience?

Revolution Backwards or Forwards?

The term "revolution", as it is mostly used nowadays, is a product of the modern age.[6] In fact, its Latin origin denotes a "return" to the beginning or simply a rotation or cycle. Linguistically, it is marked by the prefix *re* in *re-volutio* (from *revolvere*, "roll back"). When Copernicus published his seminal work *De revolutionibus orbium coelestium* – "On the revolutions of the heavenly bodies" – in 1543, this is precisely the sense in which he uses the term. The interest in the orbits of the heavenly bodies did not, however, arise from mere curiosity, but was urgent because it was considered that the cycles of the stars, while independent of all human influence, affected our lives. Astrology was then not pure astronomy but had a political meaning. Although the form of the social institutions succeeded one another, it was possible to speak, like Thomas Hobbes, of the English Revolution of 1640–1660 as a cycle from monarchy to parliamentarism, dictatorship and oligarchy, and back to monarchy. Here, the revolutionary upheaval took place within the framework of a circular movement; the concept of "revolution" in this context primarily had the supra-historical meaning of a natural process. A cyclical process, as in Turner's traditional societies.

Although it was a natural process, and certainly involved some kind of impact on society, the revolution's disruptive effect on society did not appear to be as profound as it would later be. Social

6 The following is based on Koselleck 1989, pp. 67–86, but also "Revolution, Rebellion, Aufrühr, Bürgerkrieg" 2004, pp. 653–789 and Eisenstadt 1978.

relations did not change in any radical way. As philosopher Hannah Arendt notes, "there was [no term] which would describe a change so radical that the subjects became rulers themselves".[7] But in the 17th century, it became possible to imagine that revolution, even in the form of unique political events, rebellions or civil wars, could mark a turning point in history. Previously, rebellion as a social and political category was really only about something that disrupted order, and from which a return to normality was sought. Even later, rebellion was not understood as revolution unless it led to lasting and radical change. But the so-called Glorious Revolution in England in 1688 introduced a parliamentary system that every king had to support. There was no going back, according to Arendt. With the introduction of a new constitution, history had taken a new turn. Here, the concept has freed itself from any astronomical meaning.

It was only during the Enlightenment that the concept of revolution was transformed into a historical and progressive concept that could encompass the whole of human existence. Voltaire admired the Glorious Revolution in England precisely because, unlike simple rebellions, it managed to open up a new horizon. From the French Revolution of 1789 onwards, the revolution no longer led back to the origins but forward, towards an unknown future – even if the revolutionaries themselves initially thought they were merely restoring the old order of things. Even when it was clear that the direction of the revolution was the future, it was possible to argue about whether the revolution could be accelerated or retarded; a concrete revolution might fail, and therefore be perceived as a setback, but could still, as Immanuel Kant, for example, pointed out, be a moment in the general revolution.[8] Revolution was now a philosophy of history, an inexorable force whose irreversible direction was the future and which meant a revolution of all human relationships and therefore also of human nature: "All that is solid melts into air".[9]

7 Arendt 1990, p. 41.
8 Kant 1979, p. 159.
9 Marx and Engels 2002, p. 248.

The experience of acceleration itself gave the impression that the uniqueness and irreversibility of the revolution was a break with all the past. Perhaps the most violent symbolic act to mark this break with the French Revolution was the severing of Louis XVI's head from his body on the morning of 21 January 1793. Less drastically, but perhaps just as profoundly, the break in history was marked by the introduction of a new calendar: between 1793 and 1806, the revolutionary calendar was in force, with the year of the Republic's birth as year one, twelve months, each lasting thirty days, three ten-day weeks per month and ten hours of one hundred minutes per day. Dress, too, and the very pattern of bodily behaviour, changed in two phases, first to clothes like uniforms between 1791 and 1794, and then to more free clothes that exposed the movements of the body from 1795.[10] In these and several other ways – memorials being removed and churches being turned into places for worship of reason – a radically new beginning was concretely marked. The symbolic took material form.

Revolution in this sense is, as I said, a turning point, where history takes a new and progressive direction. Both the new and the progressive are crucial. And although revolution as such exceeded the individual's capacity for initiative and power to influence, thinkers from the Romantic Friedrich Schlegel to Marx believed that human beings were not only exposed to revolution, as fate, but could also contribute to – or prevent – its progress. As Marx wrote in 1852: "Men make their own history, but they do not make it just as they please; they do not make it under circumstances chosen by themselves, but under circumstances directly encountered, given, and encountered from the past."[11]

It was literally possible to seize the moment. The possibility of planning a revolutionary social change became conceivable. The working class, or the people, were a collective subject whose timely mobilisation could accelerate the revolution. Thus, seizing power over society meant seeing oneself as an active agent acting

10 Connerton 1989, p. 10.
11 Marx 2000, p. 329.

in step with (or, in the case of counter-revolutionaries, out of step with) the rhythm and irreversible progression of history. This is what sociologist Shmuel Eisenstadt, following the Jacobins of the French Revolution, calls the "Jacobin dimension of modernity" and which he argues is a central component of the modern conception of the possibilities of politics, which is found far beyond radical political positions: that through politics it is possible to actively shape society in a radical way.[12] The epithet "professional revolutionary" became possible, but above all the idea that it was possible to control history and not just be subjected to it.

In the modern understanding of revolution, influenced by, but not limited to, Marx, a revolution is that in which everything changes. Although Arendt, among others, protests against giving the French Revolution, which was both political and social and claimed to transform everything, precedence in definition over the American Revolution, which was limited to the political, she recognises that it is the wider French Revolution that has become the pattern for what we mean by revolution, even beyond the political and social sphere.[13] After the revolution, Rome is not being re-founded, as was the political endeavour in classical times, but a new Rome is being founded, to paraphrase Arendt.[14] But what is meant by the "everything" that the revolution is said to encompass? Not only the social and the political, but also the aesthetic, the moral, the religious and knowledge itself. Revolution is a revolution of the whole world of human life and thus of man himself.

The philosopher Slavoj Žižek has repeatedly referred to a pithy quote by G. W. F. Hegel, that "it can only be considered a modern folly to change a corrupt moral system, its constitution and legislation, without changing religion, to have made a revolution

12 Eisenstadt 1999, pp. 72 f.
13 Arendt 1990, pp. 48 f. *and passim.*
14 Arendt 1990, pp. 208–212.

without a reformation".[15] What does it mean that it is not possible to make a revolution without a reformation?

According to Žižek, the problem with many revolutions is that they are not radical enough; they do not result in a cultural revolution in which the very premises of what it is to be human change. If the pre-revolutionary subject – including the subject that drives the revolution forward – is supposed to be able to live through the revolution without itself changing either during or after it, then the revolution has not been radical enough. This will then have repercussions on the success of the revolution. It is not enough to transform social relations if human beings as such are not also transformed. The revolution as a social passage involves both "objective" and "subjective" transformation; not only a new society but also a new human being. The two presuppose each other.

Žižek stands for a political version of what is known in theology as the law of the grain of wheat: that only he who first loses himself can win himself. He therefore speaks, with the psychoanalyst Jacques Lacan, of a "subjective destitution", a form of symbolic death, as a precondition for the success of a revolution. The subject acting in the revolution cannot itself stand outside the upheaval and transformation that the revolution is supposed to bring about. A true revolution concerns not only the relations of production or the laws of society but the revolutionary subject itself. Thus, there is an apocalyptic quality to this symbolic death in Žižek, when the very coordinates of what it means to be human change, with no guarantee of what life will be like on the other side of the revolution.

Žižek is certainly particularly drastic in his way of talking about this, but he refers to a dialectical reasoning that we also find in Marx and Engels. By seizing power from the bourgeoisie, the working class will ultimately abolish itself as a class and thus

15 Hegel 1979, p. 360. Cf. Žižek 2008, p. 195 and also my book Sigurdson 2012, p. 179.

class antagonisms as such.[16] From the perspective of the threshold experience, I would say that Žižek almost exclusively emphasises the discontinuity between the pre-revolutionary and the post-revolutionary state. But then the liminal transition's character of passage risks being obscured. When the pre-revolutionary and post-revolutionary states are right next to each other, the threshold between them becomes only a thin line without any density of its own. The passage becomes binary and not tripartite, as in van Gennep. For Žižek, the liminal state is a state of absolute freedom, without any particular content or personal characteristics of the subject. Thus, the turning point itself appears unfathomable and the outcome in the form of triumph or disaster as completely unpredictable.

One political criticism of such a view of revolution, as something that in all respects breaks with the past and starts from scratch, is that it seems highly risky. Whether that is a fair criticism is a question for another occasion. Here, I am particularly interested in how Žižek elucidates the subjective conditions of revolution or, in other words, what light the concept of revolution sheds on the liminal state and threshold experience. To what extent is it a total transformation of the subject? Would we recognise ourselves after the revolution? As the Marxist politician and philosopher Antonio Gramsci writes: a social crisis consists precisely in that "the old is dying and the new cannot be born; in this interregnum a great variety of morbid symptoms appear".[17] The revolution again presents us with the problem of passage: what does it mean to no longer be somewhere but not yet to have arrived somewhere else? One might ask, perhaps a little impertinently, *how* new the newness of the revolution must be for it to still be a genuine revolution. Is it enough to speak of the new *as* something new? In other words, how should the revolution as a threshold experience be understood?

16 Marx and Engels 2002, p. 244.
17 Gramsci 1992, p. 276.

The Rhetoric of Social Engineering

When the Swedish novelist Ivar Lo-Johansson publishes his autobiographical account of the 1930s, he chooses *Tröskeln* – "the threshold" – as the title of his work. *Tröskeln* refers to the political and literary upheaval that took place during this decade. Crossing a threshold in time was "like entering a room you had never been in"; in the 1920s the city streets smelled of horses and hay, but in the 1930s they smelled of asphalt.[18] Lo-Johansson describes a time of transition between old and new which, from his perspective, results in a radical change that affects both society and individuals. "That's when it all happened," as one voice says in the introduction to the novel.[19] He understands Sweden's 1930s as a threshold period, a time when we can use the concept of revolution in relation to concrete social change to deepen our understanding of the threshold.

In Sweden, there was no violent revolution in either the 19th or 20th centuries. Nevertheless, in modern times Sweden has undergone a social transformation that in some respects resembles a revolution.[20] Perhaps we can speak of a social revolution within the framework of political reformism? In the 1930s discussion of social change and housing policy, there is a rhetoric reminiscent of the revolution's emphasis on the new and progressive that is interesting to discuss in relation to the question of how new the new society really is compared to how new it claims to be. Or put differently: how should we understand the transition between the old and the new that takes place in a social change like the one Sweden underwent in the 1930s?[21] To what extent can we also see a transition manifested in the concrete material, especially in terms of the design of housing and the relationship to people? And can a binary rhetoric about the difference between the old and the new possibly conceal how the threshold between the old and the new both separates them and

18 Lo-Johansson 1982, p. 5.
19 Lo-Johansson 1982, p. 5.
20 Eisenstadt 1978, pp. 197, 206, 322.
21 I limit myself strictly to the 1930s. For a discussion on Sweden's modernisation during the 1960s and afterwards, see Wiklund 2006.

unites them? The question of whether Sweden's social change in the 1930s should be characterised as revolutionary rather than reformist is left to an examination of whether a revolutionary rhetoric at this time risks concealing the threshold character of the transition.

The 1930s was the decade of both industrialisation and urbanisation in Sweden.[22] After the recession of the 1920s, the economy grew steadily and strongly; labour market policies geared towards consensus and stability managed to combine economic growth with a developed social safety net. In a speech to the nation in 1935, King Gustaf V emphasised consensus as a long-standing Swedish characteristic, but also one that characterised modern Sweden. 1932 saw the beginning of a Social Democratic government that would not be broken until 1976. Sweden developed economically, politically and socially at a record pace between the 1930s and the 1970s, changing from a largely agrarian and not very wealthy nation to one increasingly characterised by prosperity, social justice and urbanisation. As early as 1936, the "Swedish model" became internationally recognised through American journalist Marquis Child's book *Sweden: The Middle Way*.[23]

The Swedish model – progress, consensus and centralism – is not only a description of economic and social development, but also the name of a self-understanding that was cultivated within certain intellectual groups in Sweden. The idea of social engineering, that it was possible to use rational and technical means to develop and steer the country wherever one wanted, was cultivated in a way reminiscent of a revolutionary belief in the possibility of actively steering history. Unlike the more radical revolutionary tradition, this was about "the engineer and the consumers rather than Marx and the workers".[24] The Swedish model was not just a description of Sweden's modernisation, but very much a utopia that was formulated

22 The description here is based on my book Sigurdson 2000, pp. 190–234, the references given there, and Björck 2008.

23 Childs 1936.

24 Björck 2008, p. 108, where the quotation, however, is about the starting points of US crisis policy.

as an ideal and therefore aspired to. Housing policy, in particular, was an important part of this modernisation in the 1930s. Here we encounter a radical view of the new path that the nation has now embarked on.

1930 was the year of the breakthrough of functionalism and modernism in Swedish architecture.[25] Crucial to this breakthrough was the Stockholm Exhibition, which ran from 16 May to 29 September that year. The art historian Gregor Paulsson (1889–1977) was the general commissioner and Gunnar Asplund (1885–1940) the chief architect.[26] At the time, Paulsson was one of the most ardent advocates of functionalism and Asplund one of the most renowned architects associated with functionalism. The Stockholm Exhibition wanted to present buildings that, through their modern, light and airy style, gave an idea of what a progressive future might look like. This was mainly in the field of architecture, but to some extent also in the urban environment and interior design.[27] Mass production that makes housing available to the many and not just the few was emphasised, as was industrial production of interiors. The emphasis was on the renewal of construction. They had been influenced by the European modernism of Deutscher Werkbund, Bauhaus, Le Corbusier and others. Through architecture, interior design and urban planning, they wanted to create a new human being. The Stockholm Exhibition has repeatedly been described as a revolution, and many of its supporters, including Commissioner General Paulsson, saw it as just that.[28]

---

25 See Eriksson 2001, pp. 436–497.

26 Gedin 2018 gives an overall presentation of the Stockholm Exhibition in his biography of Gregor Paulsson on pp. 115–189. As Henrik Björck points out in Björck 2004, pp. 317–322, the Stockholm Exhibition was part of a larger agricultural meeting and also had from the beginning a focus on design industry rather than architecture.

27 Cf. Johansson 1930, pp. 485–496. In fact, the art of interior design was probably more closely linked to the preceding 1920s classicism than was architecture.

28 See Gedin 2018, pp. 138, 158, 167, 193, 204, 247, 302.

In his autobiographical novel *Författaren* ("the author"), in which Ivar Lo-Johansson also dwells on the 1930s, he describes how he strolls around the exhibition, when the sun is shining and heralds a new era. He writes victoriously:

> The functionalist era had been blown in. The style of the new era was precisely the scraping off of styles. Its naked language was facts. I translated the language of architecture directly into that of literature. I went and looked around for the new human being.[29]

Lo-Johansson was far from alone in expressing such thoughts about the "new human being" in relation to the Stockholm exhibition, although he himself found it difficult to find this new human being wherever he looked. The idea of a renewal of humanity as a function of the renewal of society was expressed by the architects and art historians who were active at and around the exhibition. As Uno Åhrén, one of Paulsson's closest collaborators, put it in the same year as the exhibition, we were living in "a time when life is being rapidly reshaped" and therefore radical building solutions were needed "in accordance with our changed living conditions and our new demands". Hence the urgency of innovation: "The physical and mental health of human beings, as well as our whole cultural life, depends very much on how we organise our cities and our homes for us."[30] Or as when art critic and author Gotthard Johansson travelled abroad in 1929 to see "functionalism in action" and found himself "suddenly in a new world" where he felt for the first time that he was living "in a great, strange and innovative time". He thus became a convert to functionalism and one of the apologists for the Stockholm exhibition.[31]

Already in 1931, the year after the Stockholm Exhibition, Gregor Paulsson, Gunnar Asplund, Uno Åhrén and several of its leading architects gathered around a polemical programme document entitled

---

29 Lo-Johansson 1976, p. 5.

30 Åhrén 1930, p. 429. See also Åhrén 1925, pp. 7–36 and Åhrén 1928, pp. 173–175.

31 Johansson 1931, p. vi. Cf. pp. 14, 251–253 and *passim*.

acceptera – "accept" – in which they set out their ideas. Here, too, a qualitative difference between then and now is formulated: A-Europe is contrasted with B-Europe in the same way as "now-Sweden" with "old-Sweden"; a "new world" and a "new type of human being" with the largely obsolete culture. Only an adaptation to the new era in terms of construction could also liberate this new human being and vice versa. This is expressed in language that emphasises the distance between then and now: "An interest in the values of the time gives rise to demands for objectivity. The radical art of building is driven by the endeavour to achieve unbiased, objective design, independent of cultural loans."[32] Time itself was the driving force behind this change, but at the same time the outcome does not seem entirely certain and is described as a tug-of-war "between forwards and backwards in our culture's abrasive two-chamber system".[33] The "artistic culture" must therefore "adapt to A-Europe or become irrelevant"; "[t]hose who do not want to accept it refrain from co-operating in the development of culture".[34] This leads to the conclusion that the "present reality" must be accepted – "only in this way do we have the prospect of mastering it, of controlling it in order to change it and create a culture that is a flexible tool for life".[35] *Acceptera* has a polemical purpose: it wants to defend the functionalism that was questioned during and after the Stockholm exhibition.

Although there are some words of appreciation for tradition, *acceptera* was above all about presenting modernisation as an irresistible force with aesthetics as its tool and whose direction towards the future could possibly be slowed down but not stopped.

The radical contrast can be found elsewhere in the contemporary intellectual world, for example in the famous 1934 book on population policy written by Alva and Gunnar Myrdal. Alva was

32 Asplund et al 1980, p. 14. See the discussion on this in Björck 2008, pp. 216–223.
33 Asplund et al. 1980, p. 186.
34 Asplund et al. 1980, pp. 25, 198.
35 Asplund et al. 1980, p. 198.

a writer and politician and Gunnar an economist and sociologist. Together they formed the intellectual power couple of the Swedish 1930s. Gunnar Myrdal was also a member of the governmental committee "Bostadssociala utvecklingen" (Social Development of Housing), which was established in 1933, together with Uno Åhrén. In *Kris i befolkningsfrågan* ("The demographic crisis"), Alva and Gunnar Myrdal want to discuss the low birth rate from "objective" psychological and sociological perspectives. However, objectivity does not prevent the book's rhetoric from also offering a stark contrast between then and now: it was imperative to belong to the "woke" over those from "cultural backwaters", actively pursuing a "prophylactic social policy" by "rationalising, complicating production and the whole of human life and thus raising the quality requirements for ourselves and our fellow human beings" and thus cultivating a new "human type".[36] This actively cultivated and progressive future is something that on the one hand belongs to the inexorable course of time, but on the other hand also something that "all" people affirm – at least all those who, like Åhrén and the Myrdal couple, are themselves in step with the times.

Kris i befolkningsfrågan is one of the clearest expressions from this period of trust in the ability of social engineering to actively cultivate a new "human type" from the "human material". Rhetorically, both a strong emphasis on human agency and human passivity emerge: the elite are to lead and the people are to be moulded. Already earlier, in Gunnar Myrdal's famous article "Socialpolitikens dilemma" ("The Dilemma of Social Policy") from 1932, a sharp distinction between old and new was introduced in accordance with a new "social policy ideology" that carries "strongly radical and to some extent revolutionary possibilities", albeit not in the Marxist sense.[37] The break between then in terms of tradition and now in terms of rationality and flexibility appears to be total and the development almost like a destiny.

36 Myrdal and Myrdal 1997, pp. 10, 11, 206, 262.
37 Myrdal 1932, p. 25. Cf. Björck 2008, pp. 224–234.

If the Stockholm Exhibition was a vision for the future, the contemporary reality was poorer. In 1935–1936, Statistics Sweden conducted a census that revealed that large parts of the Swedish population suffered from poor housing conditions. This prompted journalist and author Ludvig "Lubbe" Nordström to embark on a journey around Sweden in 1938 to investigate these conditions. The trip resulted in a number of radio talks, but also a book that is usually considered emblematic of Swedish modernism, *Lort-Sverige* ("dirt-Sweden"). Nordström also strikes an apocalyptic tone when he describes the threat of physical and spiritual "dirt" that lurks in Sweden – "the world's leading cultural nation" – if modernisation is not allowed to take place and cleanliness prevail.[38] By replacing "dirt-Sweden" with "villa-Sweden" and thus the old with the new, there is a way forward, but it also involves a spiritual and moral uprising where responsibility needs to take the place of laziness.

There is a long-standing correspondence between spatiality and corporeality, according to Nordström: "The rooms were dark and gloomy and remained so. The people were dark and dirty and remained so."[39] For Nordström, everything is connected: housing standards, racial hygiene, cleanliness and spiritual health. A revolution is needed in which Sweden comes to terms with its past in order to save the country from stagnation and decay: "There is a battle in the souls!"[40] More examples of similar ideas from the Swedish 1930s could be cited.[41] The above is probably sufficient to establish that revolutionary tones recur in these writings: an almost inevitable development of history combined with an avant-garde that drives it forward (since "everyone" has rarely been able to realise the direction of history), a sudden turning point and a comprehensive nature of the upheaval, a radical change in human existence associated with the upheaval of society and, not least, the

38 Nordström 1938, p. 16.
39 Nordström 1938, p. 135.
40 Nordström 1938, p. 72.
41 See e.g. Friberger and Blomquist 1932, but also several of the articles from *Byggmästaren* referred to in my footnotes.

contrastive difference between then and now. Revolution is also sometimes spoken of, rarely in any distinctly Marxist sense, but rather as an effect of rational modernisation.

What I want to emphasise in these examples is the social engineering rhetoric of radical social change that we encounter here and which Ivar Lo-Johansson describes as a threshold time. What unites the authors of *acceptera* with the Myrdal couple and Lubbe Nordström is the way they present the connection between the new society and the new human being, as well as the conviction that it is both possible and necessary to drastically reshape both society and humanity. The prerequisite for this is a strong belief in the ability of architecture to fundamentally transform human beings, a belief that is based on humanity's malleability and flexibility.

But how radically new is this new age? None of these authors or researchers say very much about *how* this radical transformation into the new human being will take place, other than through new building. Is it perhaps the case that it is mainly the rhetoric of change that takes place in the 1930s that radically distinguishes between the old and the new? Or in other words: how do they imagine the threshold between the old and the new? What kind of experience of upheaval does this threshold between the old and the new imply? In our own time, it is probably easier to see also the continuity between what the authors of the *acceptera* called "now-Sweden" and "old-Sweden" and not only the discontinuity.

The architectural historian Eva Eriksson points out that the Stockholm exhibition to some extent builds on earlier Swedish trends in construction. The fact that everyday life was placed at the centre of the architects in the exhibition is not entirely new, but continues a line from National Romanticism through Ellen Key, Karin and Carl Larsson, Carl Malmsten and others. Compared to the 1920s, many of the buildings on display at the exhibition were more new in form than in content. Even among the functionalists themselves, continuity with earlier Swedish architecture was sometimes emphasised in "the constructively honest and practically functional".[42] Standardisation in construction had already begun and, like many of the proposals that the functionalists wrote about, was not really something completely new.

A concrete example of functionalist practice is Gunnar Asplund's extension of Gothenburg City Hall.[43] The old town hall at Gustaf Adolfs torg was designed by Nicodemus Tessin the Elder, completed in 1672, raised one storey and given a new façade by the city architect Jonas Hagberg in 1814–1817, but was found to be too small towards the end of the 19th century. After a thorough investigation, a prize competition was organised, which Asplund won in 1918, and after a long process and further changes, the Asplund extension was completed in a strongly functionalist form in 1936. What gives the current town hall its characteristic expression is the juxtaposition of old *and* new: the Tuscan-inspired classicist palace *alongside* the functionalist extension. Asplund had previously intended to work further with the classicist character of the site, but the result was instead a more abstract conformity with, as the *acceptera* states, a "consideration of the surrounding architectural scale, proportions, grouping, colour".[44] The point is of course not to claim that the Asplund extension represents nothing new in relation to the older part of the building, but that its newness in relation to the older building contains both continuity and discontinuity. Here, the architect's work speaks a language other than the hyperbolic contrast between old and new. The fact that the windows in the façade of the extension are slightly asymmetrical has the effect that it is perceived as looking down on the main building in a way that is perhaps measured but also playful.

Asplund's extension is not the only modernist work in Sweden to exhibit a similar conscious contrapuntal tension between old and new, rather than a mere contrast. There are several examples around

---

42 Eriksson 2001, p. 459, cf. pp. 453, 458.

43 On the Asplund extension, see Caldenby 2010. See also Asplund 1939, pp. 157–168.

44 Asplund et al 1980, p. 190. Cf. Eriksson 2001, pp. 468 f., 493 f.; Caldenby 2010, p. 73.

this time of a soft functionalism that draws on aspects of earlier building styles rather than representing something entirely new. On a concrete level, much of early modernism in Sweden could be understood as an attempt to strike a balance between continuity with the past and innovation, rather than a radical break with tradition. More reformist than revolutionary, one might say.

If it is true that modernist practice in the 1930s was less discontinuous with the past, then what has been written about the radical difference between the past and the present could be seen as a rhetorical exaggeration aimed at favouring the perspective taken by the authors by making it appear as history's own relentless advance. In that case, it was in the rhetoric rather than in the concrete architecture that the real breach was exhibited. If you read the annual issues of the trade journal *Byggmästaren* ("the builder") – for which Åhrén was editor from 1929 to 1932 – around the time of the Stockholm Exhibition, there is a relatively extensive discussion about the *rationalisation* that was perceived as both desirable and inevitable. Occasionally, functionalism and the accompanying rationalisation are mentioned as a break with historical decorative styles.[45] But overwhelmingly, rationalisation here is about purely practical changes in working conditions, building techniques or forms of financing against the background of a merely implied rationalisation of society as such. At the same time, there are a number of programmatic articles with more politico-social visions of rationalisation, which are keen to portray rationalisation as a natural and irresistible force, arising entirely from the quest for more appropriate solutions in a technical sense and free from utopianism.[46] The practical-technical rationalisation that is the consequence of new building materials and construction methods is

45 Johansson 1933, pp. 156, 158.

46 Cf. Bergström 1929, pp. 1–3; Kärnekull 1931, pp. 65–72; Gropius 1931, pp. 149–156. In the wake of Gropius's article and in the following years 1932 and 1933, there is a whole plethora of articles on rationalisation in *Byggmästaren*. Then the enthusiasm seems to cool down a bit. On the rhetoric of rationalisation at this time, see Björck 2008, pp. 98–102, 124–140. As Björck writes on p. 134: "In the definite singular form, the expression easily appeared as a synonym for development, an almost

used as a pretext for a more extensive political-social upheaval in the name of rationalisation. This rationalisation is sometimes presented as reformist, but not infrequently as revolutionary, with an emphasis on the difference between old and new.

In fact, the future was probably more uncertain and less plannable, and many of the architects of the time were in practice, if not always in theory, more searching and less iconoclastic. Some of functionalism's public apologists, such as the journalists Gustaf Näsström and Gotthard Johansson, either wanted to emphasise how functionalism could be seen as a continuation of some of the best tendencies in Swedish history, or eventually came to play down the contrast between history's then and functionalism's now.[47] Others, like Åhrén, maintained the contrast for political-social reasons.[48] However, no less a figure than Gregor Paulsson argued in 1941 that "the time that has passed under the sign of what has come to be called functionalism" was not "the revolutionary stage of upheaval we perceived it to be".[49] By 1945, the rhetoric had softened even more: there was talk of a "virtually" continuous collaboration between artists and industry since the time before functionalism.[50] Eriksson is quite right when she states that "[t]he period around 1930 was a transitional stage, but not such a clear-cut turning point as is often imagined".[51] Rhetorically, the upheaval was portrayed as more fundamental than was actually the case in architecture. The programmatic Swedish functionalists cultivated – for a time, mind you – a penchant for hyperbole and contrasts.

If the buildings of functionalism as they were designed around 1930 and shortly afterwards do not stand for something completely

---

world-historical category that could explain almost everything – or almost nothing."

47 Näsström 1930; Johansson 1931; Johansson 1935. Cf. Torsten Frendin's critical review of Johansson's later book in Frendin 1936, pp. 111–118.

48 Åhrén 1936, pp. 172–176.

49 Paulsson 1941, p. 42.

50 Paulsson 1945, p. 4.

51 Eriksson 2001, p. 476. Cf. Elliot 1929, pp. 73–76 and also Caldenby 2023, pp. 174–204.

new in relation to what existed before, then perhaps the new human being is not quite so new either? Perhaps the construction area between the old and the new has its own density that makes the transition between them less abrupt? This is precisely what can be called a threshold: a transitional stage rather than a breaking point. We are back, as in van Gennep's work, to the question of a liminal intermediate state in its own right, where the relationship between the old and the new in terms of architecture, society and man is renegotiated. If revolutions are passages in this sense, then the very threshold between the old and the new, be it buildings or people, deserves more attention than it receives in the kind of dualistic rhetoric we find in some of modernism's predecessors. The transition – even the revolutionary transition – is a state of its own between the old and the new, a state that cannot be reduced to a two-dimensional surface or a historical point, but a threshold state with its own physical body, so to speak, that also has an extension in time. It is a state in which the old is about to pass, but the new has not yet quite arrived, and in which our state is changeable, ambiguous and vulnerable.

In this passage, the relationship between old and new is renegotiated, but if the intermediate state is a state between old and new in its own right, then the outcome is not a given either. Considering revolutions as threshold passages poses the question not only of *how* new the new really is, but also of *what* the newness of the new age consists. There is an expectation that this new is not only new but also something desirable in comparison to the old. But how can we know this if the passage in question is really a time of radical and total change? Where is any certainty that the new really exists?

A contrasting and dualistic rhetoric of the kind we have encountered presents itself, at least in principle, as looking back at history and the present from the future, especially when the change is described as fated. But such a perspective presupposes a position beyond the revolution itself, where the visionaries already represent the new, the breakthrough of which is still controversial. Thus, they see themselves as an avant-garde with a greeting from the future, a fixed point in a state characterised by change. They imagine that they are straddling the threshold, with one leg on each side. This is probably also how many of the more radical representatives of functionalism understand themselves. They occupy a position similar to that of the master of ceremonies in traditional rites of passage, the guide who does not participate in the passage but is knowledgeable and connected to the process, and whose task it is to lead those involved in the passage across the threshold. But who really knows how to cross over well, if it is a revolution, where everything is to be new and where no examples or role models can really exist?

Revolution as Passage

A political or social revolution is undoubtedly a form of passage, but a significant difference between these and the passages I studied in the previous chapter is that revolutions occur on a much larger scale.[52] The liminal state produced by revolutionary passages does not affect just one person or a limited group of people, but a mass of people produced by the transition itself. If the new society requires a new human being and, vice versa, the new human being a new society, then the people must no longer remain with the old either personally or socially; a revolution requires a reformation, as Hegel wrote.[53] For a revolution to be truly transformative, people must be mobilised in a transition that bears the hallmarks of the liminal state, that is, in which the very foundations of social as well as personal existence are changed. According to Arendt, the revolution itself appeared to the American "men of the revolution" as "the legendary hiatus between end and beginning, between a no-longer and a not-yet" and bearing the signs of the liminal state; a break in historical time, speculations about the future, leaders appearing on stage and

52 For an overview of an anthropological perspective on revolution, see Thomassen 2014, pp. 191–213.
53 Hegel 1979, p. 360.

pointing ahead, and a certain amount of arbitrariness as a preparation for a new beginning.[54]

As with all liminal states, such a transition involves both opportunity and danger. Those leading a revolution in various ways often have a natural interest in presenting this liminal state as a creative possibility rather than a danger, and do so by presenting some kind of strategy. A strategy presupposes some kind of overview of a place that the strategist masters and can control.[55] It presupposes that it is possible to straddle the threshold. Such overview and control need not mean that the strategist himself produces the revolution. It may instead be a matter of the insight into the inevitable course of history that the strategist believes having by virtue of a knowledge of economic, social or other laws. But in that case, this insight is also associated with a will to support the direction of history towards liberation. Of course, the revolutionary may from the beginning have been more of a tactician, someone who does not possess such an overview and control, but instead must react to the circumstances that exist and somehow unbalance them in order to initiate a transition. But the question is – a question I asked at the beginning of the chapter – who leads the passage during the revolution itself, when the change of transition makes us both malleable and vulnerable? When what Marx and Engels in *The Communist Manifesto* called "the sorcerer" – in the form of the "bourgeois society with its relations of production, of exchange, and of property" – "is no longer able to control the powers of the nether world which he has called up by his spells", they still leave the question of who can take control of the process fairly open.[56] But the question remains: when the crowd gathers in the streets and squares of the city or the village green because it has had enough, when the Bastille is stormed, who will channel the revolutionary energy and give it a direction that leads to change and not just revolt?

54 Arendt 1990, p. 205.
55 See de Certeau 1988, pp. 29–39.
56 Marx and Engels 2002, p. 225.

In the traditional passage, there is both a rite and a master of ceremonies who leads the transition. It is also surrounded by a community that remains intact during the rite itself. Revolutions can be ritualised enough – like the recurring custom of gathering in the squares – but unlike traditional passages, they have no obvious master of ceremonies, simply because no one has gone through the same passage before; revolutions, in the modern understanding, are not cyclical but progressive. Having experienced a revolution elsewhere does not help much, as each revolution is dependent on its context and therefore different. Moreover, if we think of revolution as the emergence of something genuinely new, then that future is inherently unknowable. A future that is merely an extension of the past and of the present is not qualitatively new, and thus there has hardly been a revolution. Finally, unlike the traditional passages, there is no fixed time when a revolution is over.

In some kinds of almost paradoxical talk about a "permanent revolution" – a revolution without an end point? – it is not even certain that there is a real end to the revolution, in which case those subject to it remain in a permanent liminality. If those who lead the revolution by virtue of their insight into the future do not in fact possess such a capacity, partly because uncertainty and uncontrollability characterise the liminal state as such, they cannot therefore assume the role of masters of ceremonies. What role are they then playing? That of the sorcerer? Their authority must, by definition, be of a different kind than that based on knowledge, since it is impossible to have any knowledge of a qualitatively different future.

Here we can draw on sociologist Max Weber's three ideal types of political authority: traditional, bureaucratic and charismatic. Traditional or patriarchal authority derives its authority from custom, the way a political community has always been traditionally governed. It is often this kind of authority that a revolution does away with. Bureaucratic authority, on the other hand, characterises the post-traditional society and comes from education, competence and a regular appointment procedure. Bureaucratic authority invokes reason, and it is often such authority that is invoked by revolutionaries; it is by virtue of their rational insight into the course of history that they exercise their authority. But in liminal states of uncertainty and anxiety – "moments of distress – whether psychic, physical, economic, ethical, religious, political" – it is, according to Weber, a third kind of authority that can lead and govern, namely a charismatic authority.[57] The charismatic leader is the one who succeeds in capturing the imagination of his people and is thus conferred his authority as the result of an immediate and emotional relationship with the people. Charismatic authority is therefore both a personal and a revolutionary authority that does not rely on any external laws and regulations. Charismatic authority is the authority of the person who, in times of social crisis, manages to be the adult in the situation and finds a solution to a desperate situation

The word "charismatic" comes from the Greek *charis*, which can mean "grace" as well as "gift" and "favour". Weber derives the term primarily from religious leadership, which by virtue of a "gift of grace" – *charis* – could assume a leadership role, but applies it to a political context. The charismatic leader is simply the leader who exists in a liminal state, where the accepted authorities – whether traditional or bureaucratic – are no longer able to offer any guidance. Both of these other forms of authority presuppose a state of everydayness, order and permanence. Weber, by his own admission, uses the term "charisma" in an entirely non-judgemental way: it is about describing a particular kind of leadership in a particular situation, with an emphasis on the extraordinary. Whether this is the leadership of St Francis of Assisi, Long John Silver or Napoleon does not seem to matter conceptually.

But can the charismatic leader really be trusted? Interestingly, during Sweden's threshold period in the 1930s, the Prime Minister Per Albin Hanson was ascribed the epithet "father of the nation", almost a traditional patriarchal epithet signalling a relationship of trust between leader and people.[58] But how do we know that the charismatic leader is really emotionally connected to his people

57 Weber 1978, pp. 1111 f.
58 See Balázs 2020, pp. 197–200.

and not just trying to exploit the situation? How do we know that we will get through a liminal situation of societal magnitude well? As a critical extension of Weber's concept of charisma, a concept that was never actually defined by him in any final form, political anthropology has referred to the figure known in anthropology as the trickster.

According to Paul Radin, who can be said to have introduced the concept into anthropology, the trickster is "at one and the same time creator and destroyer, giver and negator, he who dupes others and who is always duped himself", someone who "possesses no values, moral or social, is at the mercy of his passions and appetites, yet through [whom] all values come into being".[59] When normality is disrupted and uncertainty spreads, there is ample room, so to speak, for attempts to mimic the charismatic authority of a leader integrated into himself and his people, but in fact manipulate the situation for the trickster's own ends.[60] The role of the trickster is to subvert and question social order and established values. The trickster thrives precisely in the in-between state of liminality, where everything is uncertain, because it becomes possible to exploit people's propensity to mimic whoever carries a strong enough message about how to survive a crisis and find a new form.

It becomes possible here to mould one's successors in one's own image. The 20th century bears quite clear witness to these possibilities through Benito Mussolini, Adolf Hitler and Joseph Stalin, but also in our own century with Vladimir Putin, Boris Johnson and Donald Trump, to name just a few of the most infamous examples. The trickster is the one who claims to rescue us from chaos, but who in fact lives off the liminal state and therefore has no real interest in seeing it end. The trickster has no substance in himself, so to speak, but lives off others, like a black hole that attracts and devours the energy of others. Therefore, since the trickster has no substance of his own, it is also possible to imitate someone else in a pretended way.

---

59 Radin 1972, p. xxiii.

60 Horvath and Szakolczai 2020; Horvath 2013, pp. 121–129. See also the various chapters in Horvath, Szakolczai and Marangudakis 2020.

The trickster can therefore just as easily be the cause of the chaos he claims to be able to remedy. The trickster, one might say, poses as a master of ceremonies, but is not in fact emotionally connected to the successful outcome of the passage. Rather, the passage serves as an opportunity for the trickster to allow himself to be reflected in the world around it by maintaining the liminal state from which he claims to be able to save us. The trickster is a liminal creature that lives on the threshold, where we are most vulnerable, and has an interest in creating liminal states where they do not spontaneously appear by presenting the present as a time of rupture.

If the trickster lives not only in but also of the liminal state, then there is no interest in it ending. In Marxist theory, but also in other political theories, the idea of a "permanent revolution" has returned.[61] For Lev Trotsky, the permanent revolution was about the fact that a socialist revolution is possible even in a country that has not yet reached the stage of advanced capitalism; simply put, it does not have to take the detour of a bourgeois revolution. It has been called a "historical-philosophical concept of compensation" to deal with the fact that revolutionary expectations have been dashed by referring to a future fulfilment.[62]

In anthropology, the term permanent revolution has been used in a similar way to "permanent liminality" to refer to an intermediate state or passage that never reaches an end, but leaves those affected by it in a state of unease and uncertainty.[63] The liminal state is characterised by both opportunity and danger. A permanent revolution, like a permanent liminality, means never being able to reach any kind of normality and stability. The permanent revolution is reminiscent of the liminoid rather than the liminal state. As the political scientist Richard Sakwa writes: "Liminality of transformation has given way to liminality of change without meaning, purpose, or direction."[64] If a revolution has a genuine emancipatory meaning, then it also intends

61 "Revolution, Rebellion, Insurrection, Civil War" 2004, pp. 761–766.
62 "Revolution, Rebellion, Insurrection, Civil War" 2004, p. 763.
63 Thomassen 2014, pp. 93 f., 208–213.
64 Sakwa 2015, p. 211.

to bring about a permanent, structural change, a new normality. But to the extent that the change does not reach its goal but remains in constant flux, then the world becomes fluid and fragmented and the subject vulnerable to the lure of the trickster. The permanent revolution becomes, in the words of anthropologist Marius Ion Bentas, "like a rite of passage that was never concluded, a failed conversion, a fake transformation, a state of not being in order, or a constant waiting".[65] Here is an environment where the double game of the trickster can thrive.

Given that no real master of ceremonies is possible, because the future is unknowable and no one can know what it holds, can there be anything else that functions similarly in a time of radical upheaval? Or are we at the mercy of whatever trickster manages to seize power at the moment? Can we even think of getting through a revolution, or is every revolution a gamble, not only personally but also socially? In order to begin to answer these questions – an answer that can only be incomplete here – we can return to the earlier discussion of how new the new really is, or how continuity and discontinuity relate to each other in such a situation.

In the example of Sweden in the 1930s, we find a crucial difference between the revolutionary, almost apocalyptic rhetoric that we saw from several leading intellectuals and the actual building practice of, for example, Gunnar Asplund, an architect who was anything but a programmatic functionalist.[66] The new functional style of the 1930s was not, as was sometimes claimed by both its critics and its supporters, an absolute break with earlier architecture, although it was in a sense something new. Asplund was an architect who carried a sensibility for earlier building. As Eva Eriksson writes of Asplund, he "lived in a time when modernity did not exclude deep experiences of the legacy of the past".[67] This is an elegant way of formulating how an orientation towards the future does not exclude or compete

---

65 Benta 2020, p. 216.
66 Cf. Asplund 1936, pp. 166–171, where Asplund distances himself from the view that form is a function of technology.
67 Eriksson 2010, p. 6.

with roots in a collective memory anchored in monuments as well as documents. Instead of a radical distinction between "old-Sweden" and "now-Sweden" of the kind established in *acceptera*, in which Asplund certainly participated, a continuity is established here in the midst of all discontinuity. Eriksson speaks of the period as a "transitional stage" but not a "breaking point", a stage with a "web of different time layers" rather than "the pure types that posterity likes to look for" when "[f]unctionalism and classicism were intertwined".[68] It is possible, so to speak, to recognise both the old place in the new and the old human being in the new. Change has taken place here through a threshold that not only separates old and new but also connects them.

Such a view of radical change as something that does not by definition preclude the existence of continuity as well as discontinuity does not mean that the new cannot be new.[69] If, for the sake of illustration, we compare it with a more individual passage, it would be as misleading to claim that there is nothing of the child in the adult after the passage of adolescence as it would be to claim that no change at all has taken place. What the successful passage here means is that the personality traits that the child possessed acquire their meaning in a new context; only in the adult do we retrospectively find what the child carried as its future. In a social context, we speak rather of collective memory, but this collective memory is carried by individuals, not least those who, like Asplund, have incorporated it into their practice. Such individuals carry a skill that is broader than cognitive knowledge, although it includes this too, in that it also includes personal experience and acquired qualities. In relation to Weber's concept of charisma, it is not just about *charis* as an individual gift, but at least as much about *charis* as grace in the sense of developing and cultivating oneself in harmony with the environment.[70]

68 Eriksson 2001, p. 476, cf. p. 491.
69 See Arendt 1990, pp. 215–281.
70 Horvath 2013, pp. 127 f.

The French Revolution can be seen as a kind of collective act of remembrance, as its rejection of the former monarchy indirectly invoked it. In order for the ritual dethronement of Louis XVI to be understood as a final dethronement of all kings and not just this particular king, it was necessary to refer to the previous royal ritual. The anthropologist Paul Connerton, speaking of collective memory, writes: "A rite revoking an institution only makes sense by invertedly recalling the other rites that hitherto confirmed that institution."[71] If these had not been recalled, for example because the memory of all the past had been wiped out, the dethronement would have been literally incomprehensible. The erasure of all memory – by banning the teaching of history, or by wiping out all traces of the past settlement – is a hallmark of a totalitarian regime.[72] Without collective memories, in the form of buildings, ceremonies or calendars, the historical sediment that enables us to produce a coherent collective narrative of a before and after would disappear. Buildings store collective memories. This is also why public buildings, but also the architectural style of a society as such, are of central interest to revolutionaries: they literally want to demolish to make way for a new human being, with different memories and different ways of moving. If form and function are linked, then a different form can lead to a different function and further to a different person.

In revolutions, as in other passages, the problem of transition or threshold is vital. It seems part of our nature as human beings that in a change we cannot oversee or control, we seek some hope of coming through well; that, although we have fundamentally changed, we can also in some reasonable sense be said to remain the same and not just become someone quite different; that those, finally, who lead the transition should not exploit our vulnerability to their advantage and our detriment.[73] Revolutions expose on a collective level the function of the threshold as a transition from one

71 Connerton 1989, p. 9. The classic in this context is Halbwachs 1968. See also Velicu 2010.
72 Connerton 1989, p. 14.
73 See Lear 2008.

state to another. Liminal periods are essential to the transformative experiences of individuals or communities. This includes how we conceive of historical change: as a homogeneous, smoothly flowing process that progresses from one stage to another, or as a more erratic development, which may also involve lostness, wrong turns and dead ends, involving the work of understanding, acceptance, forgiveness and amnesty.[74] But revolutions are not necessarily experienced as emancipatory or meaningful, even in retrospect, but can also have a traumatising effect. How should we understand a drastic restructuring of the human being on the threshold? What does it mean to experience a threshold?

74 See Benjamin 1991b, pp. 177–188.

Experiences

Passages and other transitions, such as various kinds of revolutions, involve a restructuring of the very existence of the person experiencing them. Otherwise, we would not be talking about passages, where an intermediate state is passed, or revolutions, where society is changed by a revolution. But what does it mean to experience such a restructuring? How should we understand the experience of crossing a threshold? If threshold experiences take place between the ritual and the revolutionary, they also have a personal aspect. In religious conversions, in psychological therapies and in existential limit situations, we are confronted with thresholds that can shed light on threshold experiences in general, including those that take place in the ritual and the revolutionary. With these as a starting point, I will turn here to the question of the threshold experience: the experience of crossing a threshold, but also the experience as a threshold. Why is the liminal state both opportunity and danger? To understand why this is so, we need to understand what it means to experience a threshold.

Conversions

The conversion of the Apostle Paul on the road to Damascus, as recounted in Acts 9:1–9, became, for obvious reasons, a model for how conversion in general has been imagined throughout history. As such, it is also a model of how we can imagine the nature of threshold experiences. Luke, the author of the Acts of the Apostles, tells us how Paul – or Saul of Tarsus as he was then called – persecuted the Christians. In the scene that concerns us here, he is on his way to seek out and expose Christians in the synagogues of the Syrian city of Damascus. On the way, a "light from heaven flashed around him" and knocks him to the ground. He hears Jesus' voice, urging him to get up and continue his journey to Damascus where he will receive further instructions. However, he has been blinded by the light and can no longer see, so his travelling companions have to lead him there. Once in Damascus, he has to recover: "For three days he was without sight and neither ate nor drank." Then he meets the disciple Ananias (verses 10–25) who lays hands on him, and with a greeting from Jesus, tells him that he will get his sight back and a new mission. He is baptised, eats to regain his strength and begins to preach the Gospel. This is the beginning of his mission as an apostle to the Gentile nations, as described in Acts. Paul himself tells about his conversion in the letter to the Galatians (1:10–2:10), but mentions very few details about the event.

The conversion on the road to Damascus has become a model story that has not only influenced the conversion experiences of individuals but has also left its mark on literature, art, music and so on. In the church of Santa Maria del Popolo in Rome, Caravaggio's famous painting of St Paul's conversion on the road to Damascus hangs on the right in a narrow side chapel. In this painting, Paul lies on the ground, fallen from his horse, blind and with his arms stretched upwards as a plea for help, while a bright light floods from above. It is a dramatic scene, enhanced by the play of light and dark in Caravaggio's chiaroscuro. Paul cannot see because he is blinded by the light, but at the same time he has gone from darkness to light in his view of Christians; it is as if the subject itself is calling for a chiaroscuro. The very expression "the road to Damascus" has come to signify a dramatic and radical conversion far beyond the religious sphere; having travelled this road, at least symbolically, one is no longer the same as before.

Given the importance of the conversion story, let us approach the details of Luke's description of the conversion as a phenomenon but bracket the question of what caused it.[1] The story involves sudden visual and auditory impressions that put Paul in a state of physical helplessness, a state that seems to last (a little over) three days. Paul does not regain his sight and ability to act until Ananias, on behalf of Jesus, gives him a new mission. The person who previously persecuted the Christians will now be the one preaching the Christian message. Notice that this can hardly be a conversion from Judaism to Christianity; such fixed and distinct religious identities hardly existed in Paul's time. Instead, what we have here is a modulation of the Jewish message.[2] It is clear that it is within the framework of the Jewish synagogues that Paul both persecutes the Christians and eventually preaches the Christian message. However, this is both a change of mind and, above all, a change in the mission to which Paul feels called. The account in Acts is anything but introspective; we do not witness Paul's struggle of the soul, nor do his own letters give any real autobiographical account of what Paul experiences at conversion. It is also interesting to note that it is for three days that Paul is in a state where he cannot see, cannot eat and cannot drink. Apart from the fact that the number three may have a symbolic meaning, this shows that the conversion, despite its dramatic character, is not instantaneous.

The three days correspond to the intermediate state that we recognise from the description of passages in previous chapters. The dazzling light and the voice on the road to Damascus mark a boundary between who Paul was before and the intermediate state, while Ananias' visit to Paul in Damascus marks the far boundary of the intermediate state to Paul's work as an apostle. Thus, the new mission was not given directly on the road to Damascus. On the contrary, Paul is invited by Jesus to get up and go to Damascus to receive the new mission. The intermediate state involves a road, both literally and symbolically, which is significant for my perspective on spatiality here: the road is not only a metaphor for

1 An exegetical account of the passage can be found in Pesch 1986, pp. 296–309.

2 See Zetterholm 2009.

Paul's conversion, but a symbol, because it is a particular road to a particular place. Although Paul is accompanied by his travelling companions to Damascus, this journey is not a mere transport but a wayfaring. That the road is not defined exclusively by its destination but also understood as part of the change that Paul undergoes is suggested by the fact that in this particular context in Acts, Luke chooses to speak of the Christian movement as precisely "the Way" (not just in a figurative way: *he hodos*; verse 2). In order to become who he was called to be, Paul must first travel a road, both figuratively and literally. The change he undergoes is dramatic, but not instantaneous and not binary. This is where Caravaggio's clairvoyance can deceive us.

It should be clear, if we follow van Gennep, that the road together with the three days depicts a liminal state, that is, a state of transformative character characterised by disorder and therefore both opportunity and danger. Although we have no introspective information about what is going on in Paul's thoughts and emotional life, it is clear from Luke's account that this is an experience of change of an existentially crucial nature. After the intermediate state, Paul's life takes a different direction, which is expressed in his different actions. Would it therefore be correct to speak of Paul's conversion as a passage? There is hardly any doubt that it is a passage, but like many other biblical accounts of similar events, it is not a rite in the sense of a repetition of a specific pattern of behaviour. The call Paul receives from Jesus in the Acts of the Apostles to become a missionary to the Gentile peoples is a unique calling for him. Like a revolution, it is both irreversible and forward-looking. The rite in this context is the baptism he undergoes, as a confirmation of the conversion he has experienced and the new relationship with others he now stands in. It is possible to argue that there is a pattern in that the personal divine address recurs in biblical narratives where passages of an existentially decisive nature are involved.

An example of this is found in the *Akedah*, the story of Abraham binding his son Isaac to sacrifice him to God (Genesis 22), but being prevented by God at the last minute; here also a dramatic change takes place thanks to a direct divine address.[3] But to the same limited extent as the *Akedah*, Paul's conversion is a ritual custom, since the emphasis is on the unique, personalised address and the event as an act of faith. Although Paul's conversion (more so than in the *Akedah*) has come to represent a pattern of how conversion has played out in the Christian tradition – compare Augustine's account of his conversion in his *Confessions* – this is merely a repetition in form: Augustine's conversion is not in substance a repetition of Paul's conversion, but his own unique conversion, including a unique, personal address by God.[4] Admittedly, Augustine is explicitly inspired by Abba Antony, an Egyptian monk whose life has been recounted by the third century theologian Athanasius of Alexandria, so there is reason to think that Paul's conversion gave rise to a pattern of how conversions came to be thought of.[5] Nevertheless, all these conversions are different from the rite of passage in the strict sense, because they are not ritual acts whose point is that they follow an already established pattern; however similar one conversion may be to another, the point is in the uniqueness, the personalisation.

Conversions are, if you like, linear, not cyclical. Paul's conversion on the road to Damascus is a unique event in his life. Earlier he persecuted Christians, then he was called as an apostle and preached the Christian message. Although some repetition of a conversion is conceivable, as in the Christian tradition of daily conversion where the original conversion is recalled in the mind, such repetition takes place in the context of the unique experience.

A further aspect of Paul's conversion on the road to Damascus that is fairly obvious but deserves to be emphasised is that it is not a self-induced conversion. Neither in Acts nor in Paul's own account in Galatians does Paul suggest that this conversion was something he himself sought. He finds himself afflicted by it; it is something he receives completely unprepared. Unlike a rite of passage, whose

3 See the discussion on *Akedah* and on the difference between a repeated and a unique event in Eliade 1965, Chapter 3.
4 Augustine 2006, pp. 133–154, book 8.
5 See Athanasius of Alexandria 1978.

cyclical return means that the recipient is prepared for it to come whether the passage is welcome or not, conversion for Paul has the character of a surprise. Paul *receives* the call, and his concrete conversion is the response to its invitation. This does not mean that Paul's subjectivity is reduced to mere passivity. The mode of conversion is mediopassive. He is active in the way he receives this surprise, not only in saying yes or no to it but also in the way, in action, of receiving it, in which there is a rich register of possible responses just as in every such experience.

However, there is a feature of Paul's conversion – as in the story of the *Akedah* – that recalls passage as van Gennep describes it, namely that God becomes the equivalent of a master of ceremonies in Luke's account. What Paul experiences in Jesus' address is a call from God. Not only does God appear in this context as the ultimate credible master of ceremonies who can lead the one passing through the liminal state, but the conversion itself is defined by faith in the sense of trust. "Faith" in this sense is not only the end of conversion, its final cause, but also the operative cause that gives rise to the change. In psychoanalytic language, we could speak of conversion as transference. For it to succeed, there must be an external agency of a kind we trust to guide us well through the transition's state of disorder: opportunity and danger becomes a precondition for genuine change.[6] The call of Jesus provides Paul not only with a call to a changed mission, but also with a trust that allows him to change in this direction without losing himself. He does suffer a symbolic death – in his own words, he becomes a "new creation" (2 Corinthians 5:17) – but beyond this death, it is still he, Paul, who receives a new mission. There is continuity in discontinuity.

The perspective of my account of Paul's conversion here is not theological but rather phenomenological. There are, of course, many theological questions we can ask ourselves about this conversion. What was its nature? Is it not possible to give a naturalistic explanation of the conversion and, if so, would this compete with a

6 See my discussion of trust under liminal conditions in relation to the figure of St Christopher in Sigurdson 2025b.

theological one? How does conversion relate to baptism, which has an obvious ritual character symbolising the passage to new life? And was Luke's account of Paul's conversion, despite its pattern-making character, really intended to create a pattern for future conversions? My interest in Paul's conversion here, however, is limited to the fact that it is both a clear and well-known example of what I would call a threshold experience. With the help of such a dramatic story, we can begin to see what a threshold experience involves. So it makes no difference to my argument here what Paul converts from or to. As well as conversion, I could speak of apostasy, that is, apostasy from something, understood as a neutral concept; one person's conversion is another person's apostasy. It is the *how* of conversion that is interesting in this context.

I just argued that God acts as a master of ceremonies in Paul's conversion, that is, the agency on which Paul relies in the passage to get through successfully, and thus claimed that the conversion experience involves a transference relationship. Is it possible to speak of psychoanalysis as a passage in a way that sheds light on the threshold experience? Therapy – the word comes from the Greek *therapeuein* meaning to cure – is also, like conversion, about change.

Transference

In a late text on the technique of psychoanalysis, Sigmund Freud explains that the prerequisite for a successful analysis is that the analyst and analysand first enter into an agreement.[7] This agreement involves the analysand promising to tell everything, while the analyst takes a vow of silence and promises to use his or her knowledge to understand what is troubling the analysand. The goal of the analysis is for the analysand's ego to regain control over his or her life – "mastery over lost provinces of his mental life" as Freud puts it – but this quest for regained agency is achieved, paradoxically, by

7 Freud 1981a, pp. 172–182. See also the further discussion on transference in Lacan 1998, pp. 123–130, 230–243 *et passim* and Lear 2010, pp. 117–143.

relinquishing mastery and control when the analyst interprets what the analysand himself neither understands nor controls, his own unconscious.[8] Therefore, the analysand must not only tell the analyst what he or she knows but does not want to tell anyone else, but also say everything that comes to mind, no matter how embarrassing, stupid or uninteresting it may seem. Thus, there is a chance that the analysand expresses something that he or she is not aware of because it is something pre-conscious or unconscious. The analysand has to let go of his/her critical faculties and thus relinquish his or her own self-control. Through free associations, the person can, for a moment, overcome his or her neurotic blocks. In the analysis, the analysand thus hands over the interpretative initiative to the analyst for a while.

If the analysand manages to speak freely in front of the analyst, who then also has the opportunity to interpret what he or she says, what Freud calls transference occurs. For the analysand, the analyst becomes, in Jacques Lacan's words, "the subject who is supposed to know" and who holds the key to the analysand's suffering.[9] In Freud's own words: the analyst no longer appears as someone the analysand pays for their services, like a "guide on a difficult mountain climb", but rather the analysand comes to regard the analyst as "some important figure out of his childhood or past".[10] The analysand transfers feelings and reactions to the analyst as if this person were a central figure from the analysand's past life. The analyst takes the place of the father or mother or other significant person during treatment when the analysand repeats a reminiscence from his/her psychic life as if it were applicable to a present relationship.

Such a transference can have both positive and negative consequences for the treatment. On the one hand, it can give the analyst important knowledge about aspects of the relationship with the parents that the analysand has been unwilling or unable to talk about; it can energise the treatment because the analysand becomes

8 Freud 1981a, p. 173.
9 Lacan 1998, pp. 232, 233.
10 Freud 1981a, p. 174.

anxious to satisfy the analyst; finally – and perhaps most importantly – it also gives the analyst the opportunity to understand and perhaps even revise the ideals and norms that have come to govern the analysand's life. This last possibility, as Freud points out, must be used very carefully, since the aim of the treatment is not to replace a previous dependency with a new one.[11] The attachment to the analyst is meant to be temporary; the ultimate purpose of the analysis is to restore, as far as possible, the analysand's own agency.

On the other hand, there are also (further) obvious risks in the transference, as it resembles a parental relationship. The analysand may adopt an ambivalent or hostile attitude towards the analyst, as if he/she were a child again, which may mean that he/she also resists the treatment. The analysand may also confuse the experiences of transference between him/herself and the analyst with experiences that have a bearing outside the analysis. The analyst's task is to enlighten the analysand about the true nature of transference, and then it can become an influential instrument in further treatment. The analysand's relationship with the analyst can be characterised by trust as well as resistance, but without a basic trust it is unlikely that any successful analysis in the sense of a decisive change in the analysand's subjectivity can take place. The transference helps the analysand to present his/her suffering to the analyst's interpretation.

Mostly Freud describes the role of the analyst in the treatment in a way that can be likened to that of a skilful guide: "The struggle which develops, if we gain our end and can induce the ego to overcome its resistances, is carried through under our direction and with our assistance."[12] This is the most arduous part of the treatment, overcoming the resistance of the analysand, but it is also one of the most valuable, as it can contribute to change in the analysand.[13] The mountain that the analyst has to guide the analysand over is, on the one hand, a mountain that is partly unknown to the analyst, since it

---

11 Freud 1981a, p. 175.

12 Freud 1981a, p. 179.

13 Freud was not always so optimistic about the possibility of change, but this need not concern us here. See Freud 1981b, pp. 216–253.

concerns the psyche of the specific analysand. But on the other hand, Freud has developed a topographical description – a description based on its various locations – of the human psyche in general, which means that the experienced analyst moves in partially familiar territory, insofar as one human being can be said to resemble another human being. For the analyst, it is a matter of moving in both the unknown and the familiar in a process that the analyst has committed to through the contract.

Notice in particular how the analysis relates to place: it is a place-bound wayfaring, not a place-independent transport, and it takes place through hard work on the part of both the analysand and the analyst. In another key text on transference, Freud describes this place almost as an liminal state: "The transference thus creates an intermediate region [*ein Zwischenreich*] between illness and real life through which the transition from the one to the other is made."[14] In this quotation, transference is described not only as a relationship between analysand and analyst in which a transference takes place, but as active in the production of an intermediate region (literally, a kingdom or domain in-between) in which a passage becomes possible. This intermediate state is described by Freud as a realm of its own with, one may assume, its own topography, which it is the analyst's task to get to know. This also makes it clear how the transference can be an asset in the analysis, as an exposure of the unconscious parts of the analysand's mental world, which provides a clearer map. By repeating in this provisional intermediate state, the psychic inhibitions can also – at best – be overcome by being recalled to consciousness.

It may seem that Freud's analysis is primarily concerned with understanding a psychic landscape rather than change, that it operates in an imaginary rather than a symbolic register. But in the Lacanian analysis it becomes clear that the analysis is about a restructuring of the whole symbolic world of the analysand. By confronting the real – the lack that characterises existence; the realisation that I am not and cannot be whole – in the analysis, the analysand dies a

14 Freud 1981c, p. 154; Freud 1982, p. 214.

symbolic death, just as in the liminal stage of passage. The symbolic death is the prerequisite for a new orientation.[15] The analysand must give up something of him- or herself in the analysis, but in doing so the analysand is also given a different view of the world, not just a different way of thinking about it.

Freud's analysis can obviously be described in terms of a passage, with a preliminary, a liminal and a postliminal phase. It is during the liminal phase that the change is to take place, and it corresponds to the analysis itself. The analyst takes on the role of guide or master of ceremonies, guiding the analysand through the intermediate state of the transference, which is both an opportunity and a danger, and which the analyst has experienced, both through their formal training and through their own analysis. The analyst is thus someone who is familiar with the terrain, including its more dangerous areas. What is particularly interesting in this context is that the analyst is not simply described as an "ally" in a "civil war" or a "guide on a difficult mountain climb".[16] The transference also makes the analyst function as a surrogate parent, although Freud (for good reason) resists the analyst taking on this role. However, the transference is an important element for change to occur. From the perspective I have adopted in this part of the book, one could say that it is about the necessary trust in something external that must exist for change to occur (compare Paul's faith in the previous section). The analysis itself is, for the analysand, a way of being freed from their suffering (as far as possible), but at the same time it is also a risk the analysand takes, since it involves a change in the very coordinates of their existence. It is, to use Freud's term, a transition or passage between illness and life. In relation to the analyst, the analysand is in the same situation as the person undergoing a rite of passage; they must, at least temporarily, give up their own initiative and control for the hope of being able to change. The analysand must be able to trust the analyst not to abuse their position but to honour the agreement made.

15 I touch on this theme in more detail in Sigurdson 2013. See also Jansson 2022, pp. 136–147.

16 Freud 1981a, pp. 173, 174.

In his explanations of the technique of psychoanalysis, Freud has provided a detailed description of a phenomenon that also, albeit obviously in different ways, has a bearing on processes of change other than analysis as such. If one's own analysis can be described as a threshold to be crossed, then Freud contributes, at least indirectly, insights into what it means to cross a threshold in general. Interestingly, Freud provides a topographical metaphor for the analysis in this context: the mountain climb. Conversion is like walking a road, but therapy is like climbing a mountain. What they both have in common is that they denote something that is carried out with some effort and that involves a turning point and an at least temporary loss of initiative in favour of someone else. To the extent that conversions and transference-work through therapy actually take place, we also have concrete examples of threshold experiences. What the philosopher Karl Jaspers calls limit situations can be understood as another example, an equally dramatic one, alongside these.

Limit Situations

A limit situation is an existentially acute situation, a dramatic event or realisation that has revolutionary consequences for our lives. In a limit situation, according to Karl Jaspers, one is confronted with objective and/or subjective antinomies, contradictions or paradoxes of such a nature that the existence we have taken for granted collapses before our eyes. The values we celebrate collide with each other. Death seems incomprehensible but inevitable. A pandemic or a war threatens our existence as we have known it. These antinomies have no intellectual solution derived from any philosophical worldview or religious belief, but must be existentially lived through. In this way, the world as we know it, the everyday world, loses its self-evidence. In a "representational world that distinguishes between subject and object" there is "*nothing fixed* [...], no absolute that cannot be questioned, no stop that holds up every experience and every thought".[17] For Jaspers, however, the limit situation does not mean that humans are finally running aground; instead, he chooses to liken it to a migration, one in which old houses dissolve and new ones are formed in a constant process – another topographical metaphor for change.[18] For Jaspers, this is not about specific houses, but about all the ways in which people express themselves in the world around them, by shaping the world and themselves, or in art, knowledge and philosophy. However, he understands these more as temporary stops than as permanent states; the limit situation is for Jaspers a creative impulse, where the emphasis is rather on the constantly ongoing process.

However, limit situations do not have to be about the major existential events that Jaspers himself mentions: struggle, death, chance, guilt and suffering as something invincible. Even minor events, such as moving from one home to another, can be understood as limit situations, insofar as the familiar is questioned and something new breaks in. What is crucial is that it is an event that affects the whole and not just part of the person. The person confronted by a limit situation must in some sense have become a *different* person. Jürgen Hasse writes explicitly about the move – in a concrete sense – as a limit situation and emphasises that in this case it is not just a change of location: "In the sense of antinomies, the movement here goes out of the narrowness of the only possible into the expanse of the new."[19] What makes the move a threshold experience in this sense is that the move in geographical space is accompanied by the possibilities of life in a broad sense that cease and those that are thereby opened up. The move itself, including moving out and moving in, is an intermediate state. In the move, we are no longer who we used to be, with the habits, things, rooms and relationships that characterised our life in the old home. But neither are we yet the people we will become, the people we could become, in the new one. Unlike Jaspers's limit situations, which more or less force a

17 Jaspers 2019, p. 219.
18 Jaspers 2019, p. 265.
19 Hasse 2020, p. 44.

change, however, a move does not have to be understood as forced in order to be experienced as a limit situation. There are times when people move, not because they have to but because they want to. Here, it is rather the change that forces the limit situation in the sense of an experience of antinomy.

Reorientations, like therapies, could be understood as a consequence of limit situations. Alongside institutionalised forms of these, it is possible to think of reorientation in general as a response to the experience of a limit situation. Boundary situations, especially if they include not only Jaspers's more dramatic and, to be honest, rather high-flying examples, but also more mundane ones, such as moving, illustrate that threshold experiences characterise human existence in general and in everyday life. The limit situations, in their extraordinary versions, result in change, transformation, maybe even a shedding of skin, but in their more ordinary versions they are also caused by (and cause) change. They involve some kind of turning point. As with the conversions and therapies, we find a topographical metaphor: the move.

Threshold Experiences

Beyond concrete threshold experiences such as conversions, therapies and limit experiences, is it possible to say something general about the experience of crossing a threshold? The above examples have shown that the threshold experience is not only an experience of something else, but also of becoming someone else, of a turning point in the experience of oneself. If the threshold experience is not only an experience of external thresholds, but of a threshold in the experience itself, then it should be possible to speak of thresholds in as general terms as we can speak of experiences in general.

To clarify what a threshold experience is, we must first turn to the concept of experience itself. What is an experience? To begin with, it is easy to confuse an incident (*ein Erlebnis*) that happens to us with an experience (*eine Erfahrung*). No doubt a lot of incidents happened to Paul on the road to Damascus: seeing a blinding light, hearing a voice, losing his sight, falling down, being led by his travelling companions. But these dramatic incidents only become experiences in the true sense of the word when Paul reflects on them and integrates them into his own existence by reorienting his existence in a new direction. In his own description of the event in the letter to the Galatians, Paul relates his experience of it (in very brief terms) rather than, like Luke in the Acts of the Apostles, describing the incidents that happened. For Luke, who is a direct or indirect witness to the circumstances, there is no possibility of narrating the experience in the first person, since it is Paul who has it. He therefore implies these experiences indirectly by describing the event from the outside.

In everyday language, experiences are what Aristotle referred to as "acquired skills", as something one is "practised in" or "familiar with".[20] Experiences in this sense do not consist only of sensory impressions, even if they are based on perception. Rather, experiences consist in forms of sensory impressions processed through reflection. Experience therefore becomes something that is acquired, as Aristotle argued, not something that is innate or spontaneous just by keeping the senses open. You *perceive* an incident but you *make* an experience. Acquiring experience is therefore something in which the experienced person must be involved. To be familiar with something through experience is to be involved in it in some way. To have an experience of something is therefore always to have an experience of and with oneself.

Thus, it also appears that experience as such involves the experience of a threshold. According to the philosopher Hans-Georg Gadamer, the newly gained experience, if it has indeed become an experience, constitutes the horizon for every new experience: "The experiencer has become aware of his experience; he is 'experienced'. He has acquired a new horizon within which something can become an experience for him."[21] In a way, we could say the same thing about Paul's conversion.

20 See "Erfahrung" 1972, pp. 609–617, but also Gadamer 2006, pp. 53–70, 341–355.

21 Gadamer 2006, p. 348.

Strictly speaking, therefore, one can never have the same experience twice. Experiences of this kind also involves experiencing a limit to one's own initiative and control, such as Paul encountered when he answered the call on the road to Damascus. To have an experience is not to accumulate incidents in a row, but to allow an experienced incident to transform oneself, not as a purely passive relationship to what is experienced in the incident, but again in a mediopassive mode. ”Thus experience is experience of human finitude”, Gadamer writes, thereby emphasising the limits of our human capacity for initiative.[22] Experience is, one could summarise, personal, dialectical and limit-setting. Experience as such is threshold experience.

To have an experience is, as I said, to be involved in something that adds something to our previous experience, something new and other, that makes us someone partly different from who we already were, someone a little more experienced than before. Moving is not just a change of geographical location, but an existential change involving place as well as atmosphere and habits. Having lived in another country for a while means both realising that it is possible to live differently from what is taken for granted in one's own culture and seeing one's country of origin in a new way. Experiencing a personal crisis means gaining a different perspective on one's existence. All these examples of experiences, which can of course be varied in many different ways, show that experiences have to do with the relationship between the self and the unfamiliar. But this relationship cannot be thought of simply as a relationship between two entities that exist independently of each other; it is not a dualistic opposition. If living abroad for a while or experiencing a life crisis makes one different in some sense, in both cases it is *I* who become different and I who become *different*. In the experience, the personal and the foreign are somehow intertwined in a passage. It is the realisation of how such a passage takes place that can say something general about the experience of crossing a threshold.

The first step in such a passage is to encounter something that is not me but that claims me. It may be a thing, an event, a new language,

22 Gadamer 2006, p. 351.

a person. The crucial thing is not what or who this something is that asks something of us, but where it comes from and to whom it is addressed. This something has something to do with me when it requests me; it claims me in some sense, so that my attention is drawn to it.

Gerundive denotes a verb form with an adjectival function. For example, the Latin *amandus* means ”one who should be loved”. Gerundives are also found in English as nouns and proper nouns: agenda means ”something that should be done” and Amanda ”she who should be loved”. According to the philosopher Bernhard Waldenfels, the first moment of a passage has the character of such a gerundive. He calls it an anonymous precursor to the imperative: “see!”, “hear!”, “follow me!”. Something intrudes, something appeals to me, something urges me to pay attention to that something. What characterises this first moment is that it is not I myself who is its cause. My attention is drawn to something (or it is not, but then I do not have an experience).

This something that prompts us is in some sense on the other side of the threshold, because it is different from the familiar. “As in awakening and falling asleep, we cross a threshold that separates the familiar from the unfamiliar, the visible from the invisible, the audible from the inaudible, and the touchable from the untouchable.”[23] From there it attracts us, startles us, arouses our curiosity or wonder or just irritates us. It can be something that occurs to us, a thought, a word someone says or a gesture someone makes. It doesn’t have to be big or dramatic. The main thing is that it catches our attention and that we thereby relinquish the role of being the initiator. In some cases, it can be a matter of events that force themselves on us. In the case of a car accident or a sudden gale at sea, we are put in a state where we have to react to what is happening. But the sudden creative impulse in writing an essay, the unconventional choice of colour in a van Gogh painting we are looking at, or the kind word from a stranger, have the same structure as these events, albeit in a less dramatic form. Our attention is drawn by something else.

23 Waldenfels 2018, p. 99.

If this request arouses our attention, then our reception of it involves a modulation of our subjectivity. Something happens to us. It is an event, not an act. Paul does not repent in any active sense; he repents by responding to the call. In the face of the request, we are receptive and passive rather than acting and active. To use a precise English word – which I have already used in my interpretation of van Gennep – we *undergo* something in the threshold experience. Something *happens* to us in which we are *involved*. This something pulls us along, if we let it, and thus we also lose control of the event for a while, at least insofar as we are not the ones in control. When the initiative does not come from us, this challenges the primacy of agency in us; we *react* by responding to this initiative that meets us from outside.

When we talk about being afflicted with something, we can also say, in an unusual expression these days, that we *suffer* it. In our time, suffering is mainly associated with pain, mental or physical, but pain is really only one form of suffering.[24] To suffer something means to *endure* something that is not familiar or manageable. As Waldenfels puts it: "To suffer means that something happens to us, tears us out of the familiar."[25] The Latin word *passio* – "I suffer" – recurs in the word passion, and to be passionately devoted to something means to accept that something on its own terms, even when it contradicts, challenges and changes the conditions that have hitherto been the coordinates of one's own existence. Love is a form of passion, and to love someone is to not want to be oneself without the other and thus in some sense it is to give up one's own initiative, one's autonomy. But love and pain are both forms of suffering. What they have in common is the experience of undergoing something.

In order to better understand the experience of receiving something, we can usefully analyse its components, in particular "call", "response" and the threshold between them. To begin with, when something calls us, this experience involves receiving that something. The call addresses us and asks for our response, a

24 See my discussion in Sigurdson 2019, pp. 87–100.
25 Waldenfels 2018, p. 102.

response that cannot be derived from the call itself. The call cannot be likened to a hostile invasion. Nor, therefore, are we merely passive in the face of the call, like a river hit by a rainstorm and overflowing. Call and response are delimited by a threshold that both unites and divides. This threshold determines both what will become of the call and how we will respond to it. If the call asks us a question, we respond, if only by directing our attention to the call. But, as a response, it presupposes someone who is first called upon: an ego in which the dative form "me" is primary over the nominative form "I". Our comportment, when we undergo something, can be described as mediopassive, rather than purely passive or active. Our first response to the call is through its reception, like the experience of being tickled where no substitute is possible, where no one can be tickled *for* us.[26] My reception of the call, as I said, cannot be derived from the call itself, but is in a genuine sense a creative response from myself, which presupposes that I am already someone. The call is addressed to *me* and *I* respond to it. Already here we sense a duality in the receiving subject.

The call, appeal or rapture affects us and draws us out of ourselves, out of the habits we hold on to and all the familiar things we surround ourselves with. Through the threshold experience, we become someone else than we were before. Such a change involves a break with the past. Crossing the threshold involves a moment of qualitative change, of discontinuity, and cannot be understood as a continuous process. Instead, this moment can be described as a shift in perspective. A classic example of this is the philosopher Ludwig Wittgenstein's graphic figure, which can be perceived as either a duck or a hare, but never both at the same time.[27] The moment I catch sight of the duck, the hare has disappeared from my view and vice versa. The change of perspective is typical of the passage: through it I cease to see the world from the perspective of a child and instead understand the world, but also my former childish perspective,

26 Cf. my discussion of the phenomenology of tickling in Sigurdson 2021, pp. 201–217.
27 Wittgenstein 1986, pp. 193–197.

from the adult's horizon. Similar shifts in perspective occur in other threshold experiences such as pain and love, but also in intellectual breakthroughs, learning a language, and moving house. When I learn a new language, it is only when I can understand the old language through the new one that I have truly crossed a threshold in this respect. Or, to take moving as an example, when the new place is experienced as home. The difference between the pre-liminal and the post-liminal state, to use van Gennep's terms, is a difference in the horizon of experience. In the liminal state, the shift between the horizon of the preliminary state and the horizon of the post-liminal state takes place. It is in this shift of perspective that we lose the initiative, lose control, give up ourselves in order, at best, to find ourselves again as someone else. Without this moment, real change is not possible, but this possibility also means danger, the danger of losing oneself, of not crossing the threshold well.

Once across the threshold, we experience the world in a new way, because the horizon of our experience has changed. "When I was a child, I spoke like a child, I thought like a child, I reasoned like a child," writes the Apostle Paul (1 Corinthians 13:11) in what could have been a phenomenological statement, and continues: "When I became an adult, I put an end to childish ways." Paul is not still on the threshold here, but has left it behind him. So it is with the threshold experience. The new, irritating or startling thing that once attracted our attention and, through our response, broke us out of the familiar rut, will eventually be integrated into a new horizon whose coordinates have been displaced by the other. We do not live on the threshold all the time; to be constantly prompted, irritated or surprised would be to lose the horizon of ordinariness against which the unusual stands out as such. To the extent that we can learn anything from our experiences, it presupposes a horizon of experience with a certain stability, otherwise there could be no permanence in the learning.

Being an adult is not only a *different* state from being a child, it is also, hopefully, having reached a more evolved state, a state characterised precisely by a broader and deeper horizon of experience. What was new on the other side of the threshold has been integrated into my current experience by changing me and making me a different person: the language I learnt or the other culture I acquired is now part of my repertoire; the person I got to know is now one of my friends; the accident I suffered I have learned to live with. One day, the experiences of the revolution will also become everyday life. There is a before and an after rather than an either/or. Now, as Paul writes, I have put away childish things. I have become someone else, but not necessarily someone *completely* different from before; the new experience I have had does not have to be a new experience *alongside* all the others before, but can also be an experience I have *with* all my previous experiences. When the child becomes an adult, it does not become a *different* person from before; rather, it becomes the same person in a different *way*. The child's past experiences are integrated into the adult's. Paul's conversion was not in any simple way a change of belief, but a change *of* his previous state.

Does this mean that everything that was new and other on the other side of the threshold in the postliminal state loses its newness and strangeness in favour of something familiar and unique? Not necessarily. The incorporation of the new and foreign into one's own experience can take different forms. One extreme would be to cling so tightly to the familiar that the new and other can only assert itself on the terms of the familiar, which basically means that everything stays the same. As in our metabolism, the new and unfamiliar will be digested until it is incorporated into the system. The other extreme, then, would be to escape into the extraordinary and the exotic, thus saying goodbye to what was previously one's own. Crossing the threshold involves a dialectic between the domestic and the alien, and both extremes involve setting this dialectic aside in an attempt to find a foothold either before or beyond the threshold. But the threshold will then no longer function as a threshold. It may seem to have been bridged, but at the same time it has been neutralised or avoided as a no-man's land where one's own orientation has no known coordinates to adhere to.

The threshold, as I have repeatedly emphasised, has the function of both separating and uniting; it is at once privative and copulative. To have truly crossed the threshold means to have had the experience that what is one's own borders on the alien in a permanent way. A human being on the threshold is not master of its own house. Waldenfels once again: "Alienation in its radical form means that the self is in some sense *outside itself* and that every order is surrounded by the shadow of the *extra-ordinary*."[28] There is, then, an inherent otherness in the human self that constitutes the very condition of the threshold experience. A human being is a threshold being who does not coincide with itself and who is therefore capable of threshold experiences. The experience of a threshold presupposes that I myself am a being of the threshold.

One function of the threshold experience is to reveal – in the literal sense of the word: to remove the veil – this characteristic of human existence: that our own alienation constitutes us as beings, that we really are ourselves when we are outside ourselves. It is not possible to speak of a separate identity, either in the individual or in the cultural sense, without also speaking in the same breath of relationality, because we become ourselves only in relation to something or someone who is not ourselves. The threshold is the boundary that produces the differences by both uniting and separating. The threshold experience thus involves not only an encounter with the unfamiliar or alien, but also the presentation of our own alienation to ourselves. To take a few brief examples, birth, like falling asleep and waking up, is an experience that intimately belongs to ourselves and make us who we are, but which radically eludes our own initiative and also show how alienation is rooted in ourselves. Such experiences are prime examples of threshold experiences.

To conclude: the threshold experience is an experience of something that we suffer: I am born, I am called, I am ravished, I fall ill, I fall asleep, I wake up. In the threshold experience, we realise that we are not the masters of our own house. The threshold experience is not only an encounter with something other, but also an experience with ourselves, that we are outside ourselves. However, such a loss of self as the threshold experience implies

28 Waldenfels 2018, p. 116.

does not constitute a complete alienation from one's own self, as if we were to be dissolved by alienation. In order to experience the foreign as foreign, we still need something familiar in relation to which the foreign can be foreign, even if it is our own foreignness. The experience of not being master of one's own house presupposes, figuratively speaking, a house of my own that I am somehow not the master of. The threshold experience is a dialectic between the familiar and well-known and the unfamiliar and alien.

It is not surprising that in history – and even in our time – the threshold is surrounded by various devices to protect those who cross it. The threshold is both an opportunity and a danger, because it concretises the fact that change, in a sense, implies a loss of self and at the same time exposes the alienation that is an inevitable part of our human nature. The threshold materialises the realisation that we are vulnerable. Waldenfels rightly points out that no intermediary with one foot on each side of the threshold is conceivable.[29] Such an intermediary would stand above or outside the threshold itself. This, however, distinguishes the intermediary from the master of ceremonies discussed earlier in this book, i.e. the one who is connected to the passage and who has himself been granted a threshold experience. In other words, the role of the master of ceremonies cannot be to create an overview or control by standing across the threshold, but to assist, by virtue of a previous experience, the passage across the threshold. In claiming overview or control, the trickster differs from the master of ceremonies.

If we can go back and forth across a threshold unhindered, it has ceased to be a threshold.[30] Then I am no longer transcending myself in my approach to something else; I am no longer in motion, or, to put it more emphatically: I *am* no longer motion. As philosopher Ute Guzzoni has made clear, both living and wandering belong to human existence as two basic modes of existence: to live without wandering would be to remain immobile, but to wander without living would

29 Waldenfels 2018, pp. 114 f.
30 See Waldenfels 2013, p. 9.

mean an incessant restlessness.[31] Living and wandering need each other, like inhalation and exhalation or light and darkness. The threshold experience is similar, except that it involves an irreversible change. The experiences I have had, I have had, even if I can forget or repress them. Just because it is possible to regress, this does not mean that I literally become a child again.

Conversions, therapies and borderline situations are often described through topographical metaphors: the road, the mountain climb, the move. The final question that will summarise this part of my book is whether the threshold experience also concerns the experience of the concrete, physical thresholds that we cross in homes, in houses and in cities. What function do the concrete thresholds in space have for our existence?

31 Guzzoni 2017, pp. 11, 13.

Passages, revolutions, conversions, therapies, limit situations – all are dramatic examples of existentially significant thresholds. They are all usually associated, more or less explicitly, with concrete places, buildings or roads. But is it also the other way round? Do the concrete, fairly everyday thresholds, such as the threshold to my office in the introduction, also have existential significance? Do they also symbolise an entry and a transition of some kind? If the threshold is a symbol, not just an image, we can expect that the concrete thresholds we encounter in our everyday lives bear some of the characteristics of the more dramatic examples of thresholds. What can the city's neighbourhoods, streets, buildings and dwellings teach us about the function and existential meaning of the threshold? Perhaps it is here that we encounter the threshold in its most mundane, and therefore overlooked, form?

Walter Benjamin's Thresholds

A central and recurring theme in Walter Benjamin's writings, not least in his *Arcades Project* on Paris as the capital of the 19th century, is thresholds. Benjamin is well aware of their symbolic significance as both a material and an existential phenomenon. Even if, as we have seen, Benjamin seems to claim that our time has become poor in threshold experiences, contemporary thresholds are palpably present in his work. There is a dialectical tension in Benjamin's work between the anonymity of contemporary passages and their nonetheless undeniable material existence, a tension that is manifested in the transitional markers, sometimes in almost parodic form, that are still present at the thresholds.

The existence of thresholds in Benjamin's work is characterised by an ambivalence that is found in the tension between the perceptible and the imperceptible. Since ambiguity is, as he writes, "the appearance of dialectic in images, the law of dialectics at a standstill", it is not surprising that thresholds occupy the central place they do in his work, whose very form and structure of thought is dialectical.[1] Thresholds are the concrete manifestation of the dialectic between before and after, outside and inside, or separation and union *par excellence*. They carry their ambiguity as a property of being.

Benjamin associates thresholds with other phenomena in the city that he is interested in exploring: borders, gates and passages. He offers, one could say with a term borrowed from himself, a lesson in threshold knowledge, *Schwellenkunde*. His *Arcades Project* (and also his other works) not only thematises the threshold as a phenomenon but is also based on his own experience of crossing symbolic and concrete thresholds as well as his interpretation of history as a threshold phenomenon.[2] Benjamin can guide us across the threshold to a greater realisation of the concrete thresholds we encounter in our everyday lives and their significance for our existence.

To begin with, we can note that Benjamin repeatedly speaks of the "enchantment" of the threshold – "dem Zauber der Schwelle", "Schwellenzauber".[3] Benjamin is well aware of the fact that the traditional threshold is surrounded by patron deities. He calls these "penates", the kind of household gods that in the ancient Roman Empire were considered to protect the home and thus regulate the passage over the thresholds; they are also likened by him to the Delphic oracle that with its γνωθι σεαυτον – "know thyself" –

1 Benjamin 1991, p. 1246; Benjamin 2002, p. 10. See Buck-Morss 1989.
2 See Benjamin 1991, p. 147 (C 5a, 2). See also Menninghaus 1986, pp. 26–58.
3 Benjamin 1991, pp. 141 (C 3, 5), 283 (I 1 a, 4); Benjamin 2002, pp. 88, 214.

guards the threshold.[4] The penates are both a protection for, and an exhortation to, the passer-by and indicate the fact that crossing the threshold is considered a precarious event that requires the passer-by's sincerity and attention. In *Berlin Childhood around 1900*, he speaks of "the race of threshold dwellers", those "who guard the entrance to life, or to a house".[5] Here too, in these "threshold dwellers", we are dealing with caryatids and atlantes, putti and pomonas, those demigods whose moulded or carved form is supposed to protect the entrance to a house. If the threshold is enchanted, a guardian god is also needed to watch over the person who crosses the threshold, so that they get through safely and do not perish during their vulnerable time in the intermediate state. We recognise the theme by now. The threshold is both a danger and an opportunity and we therefore need protection when we cross it.

The term "enchantment" may bring to mind a thesis widely discussed at the time about the "disenchantment" of modernity, particularly in the work of Max Weber, where its loss becomes part of the modern rationalisation of the world.[6] A disenchanted world would be a world without thresholds. For Benjamin, however, his use of "enchantment" should be linked primarily to his own concept of "aura", a kind of bodily and affective presence of a work of art in a specific place. In one of his most famous essays, as well as in the passage work, Benjamin argues that the artwork loses its aura when it can be mechanically multiplied; it loses its authentic and unique status and ability to manifest something other.[7] The aura is a kind of experience of transcendence, not necessarily of something divine but of something not immediately available to human comprehension. In the age of technological reproduction, the work of art is "disenchanted", no longer part of a ritual and no longer retaining its specificity of place, but instead becoming part of

4 Benjamin 1991, p. 283 (I 1 a, 4); Benjamin 2002, p. 214.

5 Benjamin 2006, p. 57.

6 For a critical discussion of Weber's concept of disenchantment, see Joas 2017, p. 355–417.

7 Benjamin 2007a, pp. 217–251. See also, for example, Benjamin 1991, p. 433; Benjamin 2002, p. 343.

commerce, exhibition and politics. Speaking of Charles Baudelaire, in one of the fragments of his work on passage, Benjamin compares his spleen to the "decline of the aura" and quotes *The Flowers of Evil* about the "adorable spring" that has "lost its fragrance".[8] An enchanted threshold for Benjamin would rather be a threshold that has not lost its aura, that is, the ability to generate a specific experience in the passing person of being affected in an existentially pervasive way right there and then – on the threshold.

Unlike artwork, the physical threshold can hardly lose its site-specificity: its localisation in geographical space. But has the threshold in a more qualified sense lost its enchantment for Benjamin? There are some signs that this is the case. For the Parisian flâneur – the epitome of someone who observes but does not participate – the city has been transformed into the interior of a flat, and the blocks into rooms. But unlike the ordinary rooms of a dwelling, the city can "appear to someone walking through it to be without thresholds".[9] To the stroller's eye, the space does not seem to be delimited by any qualitative differences marked by thresholds. Moreover, as I mentioned earlier, Benjamin argues that in our modern age "[w]e have grown very poor in threshold experiences", especially since the passageways have been obscured by the modern experience of life.[10] So it seems that threshold experiences, like the aura, have disappeared from people's contemporary perception of space. Is Benjamin suggesting that it is the straight boulevards built by Baron Haussmann around the middle of the 19th century in Paris, which more or less wiped out the old Paris, that have made the thresholds disappear?[11] However, in the same passage, Benjamin emphasises that there are indeed thresholds in the city that cannot be ignored in

8 Benjamin 1991, p. 433 (J 64 a, 1); Benjamin 2002, p. 343.

9 Benjamin 1991, p. 531 (M 3, 2); Benjamin 2002, p. 422, cf. p. 437 (J 66 a, 6) resp. p. 347.

10 Benjamin 1991, p. 617 (O 2, a 1); Benjamin 2002, p. 494.

11 Benjamin 1991, pp. 1234–1236; Benjamin 2002, pp. 11–13. For a work that is somewhat critical of Benjamin's view of Haussmann's regulation, see Kirkland 2013.

material terms, including between different neighbourhoods.[12] And in another text, the gaze of the flâneur is not so much an attempt of overview, but rather the flâneur is a threshold connoisseur, with a large register of sensory impressions and memories, which is probably closer to Benjamin's own experience of flâneurism.[13] The fact that Benjamin returns to his own experience of thresholds, from childhood to adulthood, also indicates that he does not regard the threshold as a symbol that has gone out of date. A world without thresholds would be chaos.

In fact, recurring fragments in his passage work show that the time of the threshold is not entirely over. In the modern experience of space, too, we encounter penates, but of a different kind. Benjamin returns to various kinds of automatons – "the hen that lays the golden praline-eggs, the machine that stamps our names on name plates, slot machines, fortunetelling devices, and above weighing devices" – that line the thresholds of more mundane gateways: "the entrance to the skating rink, to the pub, to the tennis court, to resort locations"; in another place, "in front of the arcade, the skating rink, the swimming pool, the railroad platform".[14] Sometimes these thresholds seem to be located in the city, sometimes Benjamin thinks that they belong more to the destination for excursions, but in any case, even in modernity, they retain some of their mystery. Billboards for various types of bitter aperitif seem to be able to act as penates at the various stations of the Paris metro: "Amer, Picon, Dubonnet are guardians of the threshold."[15] Furthermore, the gate has evolved from the "field of experience proper to the threshold" and "transforms whoever passes under its arch".[16] Even in the "interior of the bourgeoisie" this enchantment lives on: "Chairs beside an entrance, photographs

12 Benjamin 1991, p. 141 (C 3, 3); Benjamin 2002, p. 88.
13 See Benjamin 1988, pp. 416–421; on Benjamin's changing view of the flâneur, see Buck-Morss, 1989, pp. 304–307.
14 Benjamin 1991, p. 283 (I 1 a, 4); Benjamin 2002, p. 214; p. 138 (C 2, 4) resp. p. 86, cf. p. 141 (C 3, 4).
15 Benjamin 1991, p. 136 (C 1 a, 2); Benjamin 2002, p. 84.
16 Benjamin 1991, p. 139 (C 2 a, 3); Benjamin 2002, p. 87.

flanking a doorway, are fallen household deities".[17] The power with which the doorbell can send its appeal to us who inhabit the floor demonstrates the enduring power of the "magic of the threshold", Benjamin says: "some things shrill as they are about to cross the threshold".[18] Finally, there is also a reminiscence of the threshold in the French cards where illustrated windows and doors could be pushed aside to reveal the risqué motifs of the lithographs.

All of these examples are drawn from areas of commerce, consumption and a certain period of bourgeois life in Paris. Benjamin's approach to these representations of the threshold and the reminiscences of enchantment found in their commercialisation can of course be understood as critical: "the business world knows how to make use of the threshold".[19] Nevertheless, it means that the threshold has not disappeared in a levelling out, but remains, albeit in a parodic or commercialised form. The gates to the passages are also thresholds, which are not marked by a "paved step" but are nevertheless reflected in the experience and the consequent condition before them: "Tightly measured paces reflect the fact, altogether unknowingly, that a decision lies ahead."[20] As a phenomenon, thresholds seem to exist almost everywhere, from city blocks and various transitions between public functions to the homes of Parisians, and even there in the small details. For Benjamin, neither the reality nor the ideal is a threshold-free existence, even when it comes to Paris' commercial passages, but a dialectical approach to, and a possibility of, living with the thresholds that do exist.

In the recurrent criticism of a society in which significant experiences are becoming less and less important, we thus find a defence of the threshold. The commercial thresholds Benjamin conjures up, and the capitalist society that gave rise to them, can of course be criticised as such, but they nevertheless express a

17 Benjamin 1991, p. 283 (I 1 a, 4); Benjamin 2002, p. 214.
18 Benjamin 1991, p. 141 (C 3, 5); Benjamin 2002, p. 88; cf. p. 283 (I 1 a, 4) and 214.
19 Benjamin 1991, p. 138 (C 2, 4); Benjamin 2002, p. 86.
20 Benjamin 1991, p. 142 (C 3, 6); Benjamin 2002, p. 89.

fundamental experience of discontinuity that transcends their current form. Above all, it becomes clear that the human experience of thresholds and the experience of concrete, spatial thresholds are linked. The threshold, for Benjamin, becomes a symbol that holds together experience and spatiality. The thresholds of Paris – its borders, gates and passages – are characterised by the society they shape. But through the criticism of the thresholds that he exercises in his Passage-work, the possibility of a passage is also found through and from this figure. Space itself harbours unimagined possibilities for those who know how to critically experience it. Threshold knowledge and threshold criticism go hand in hand: getting to know thresholds in order to be able to cross them also requires setting limits to their validity and thus establishing another threshold to cross. This is precisely what Benjamin aims to do in his *Arcades Project*. His dialectical critique of the present does not imply that the thresholds are to be dissolved or replaced; instead, he likens them to awakening from a dream or, in other words, to another passage, this time between sleep and the waking state.[21] The rehabilitation of the thresholds should therefore not be understood as a remythologisation of them, but as a dialectical reminder of their experiential inevitability.

The Paris Neighbourhoods

For Benjamin, cities are experiences of diversity and difference. They are only ostensibly homogeneous, as is evident in their many boundaries: "Nowhere, unless perhaps in dreams, can the phenomenon of the boundary be experienced in a more originary way than in cities."[22] To know a city is to know where its borders are: "along railway crossings, through houses and parks, along riverbanks" as well as in "the enclaves of the various neighbourhoods". This does not mean that the boundaries are impenetrable: "The boundary

21 Cf. Benjamin 1991, p. 491 (K 1, 3); Benjamin 2002, p. 389.
22 Benjamin 1991, p. 141 (C 3, 3); Benjamin 2002, p. 88.

extends like a threshold over the streets; a new district begins like a step in the dark, as if one had stepped on an unnoticed, lower step." Elsewhere he makes a clear distinction between boundaries and thresholds: "The threshold must be carefully distinguished from the boundary. A *Schwelle* 'threshold' is a zone."[23] When Benjamin talks about borders, however, he take them to function as thresholds, as zones. Cities are crossed by various borders and these borders are experienced as thresholds. A city is not like a jigsaw puzzle where the borders mean that the pieces are close together. It is possible to cross these borders, precisely because they are thresholds, but on the other hand the passage is not frictionless. Something is different on the other side. This means that cities, according to Benjamin, uniquely offer the very threshold experiences I am interested in here, and they take place in relation to very concrete thresholds, albeit in the form of streets, railway crossings, river banks and so on.

It was in Paris, at the Bibliothèque Nationale, in the Salle Labrouste, the library's impressive reading room with its nine circular ceilings, that Benjamin wrote much of his work on the passage, and it was Paris, as I mentioned, that was the subject and example of *The Arcades Project*. Paris is divided into twenty arrondissements and each of these also into four neighbourhoods, but the historic neighbourhoods as they existed before Baron Haussmann's regulation still live on in the collective memory. Benjamin comes close to the idea that the different neighbourhoods of Paris carry their own atmosphere, and crossing the border into another neighbourhood is such a clear change of atmosphere that it deserves to be called a threshold.

Can this really be true? Do people notice that they change neighbourhoods, even if they are occasional visitors, more or less unaware of the administrative or historical boundaries? Let's take the border between the Marais and the Bastille as an example. According to French writer, editor and medical doctor Eric Hazan, the border is distinct: "If you cross Boulevard Beaumarchais and turn down Rue Amelot, you are conscious of leaving the Marais

23 Benjamin 1991, p. 617 (O 2 a, 1); Benjamin 2002, p. 494.

for the Bastille *quartier*."[24] Boulevard Beaumarchais, as the name boulevard suggests, is a wide, straight, busy street; Rue Amelot, just inside of it, running obliquely parallel to the boulevard, is a long, straight, less busy street. If we compare it with the Marais' maze of small streets, there is, also in a purely physical sense, a considerably different character to the urban plan we encounter in the Bastille. It is obvious that this is so, even to a very occasional visitor to Paris like myself. Crossing the Boulevard Beaumarchais means changing atmosphere.

In a way, it is a historical coincidence that has helped to create this difference. Haussmann had intended to continue his demolition of historic Paris by extending Rue Étienne-Marcel all the way to Boulevard Beaumarchais, which would have meant crossing right through the Marais. Rue Étienne-Marcel is a four-lane, busy road flanked by shops that crosses the first and second arrondissements. When it reaches Boulevard de Sébastopol, it briefly changes its name to Rue aux Ors and then to Rue du Grenier-Saint-Lazare, but then it stops at Rue Beauborg. For some reason, it was not extended to Boulevard Beaumarchais, with the result that much of the historic Marais remains. Instead, another street, Rue Michel le Comte, leads into the Marais from Rue Beauborg. Rue Michel le Comte is a virtually single-lane street, with almost no car traffic, lined with beige sandstone façades. Many of these probably conceal residential buildings. The bustle and traffic of Rue Étienne-Marcel is not noticeable. Of course, there are busy streets in the Marais – I am making a note of these impressions in a café along the street into which Rue Michel le Comte opens, Rue des Archives, with two lanes of traffic almost in front of my face – but it is no exaggeration to say that there is a different pace, a different rhythm to the Marais, thanks largely to the older, seemingly more arbitrary configuration of its streets. The Marais is, one might say, part of the old Paris, a neighbourhood where it is still possible to get lost. Therefore, it is also possible to experience a change of atmosphere when walking

24 Hazan 2011, p. 3. Hazan 2021 also offers many insights into Paris.

into the Marais via Rue Étienne-Marcel from Les Halles or, for that matter, from the Bastille via Boulevard Beaumarchais.

However, the Marais is not just a series of streets and buildings, but has a history of its own. Although nowadays the neighbourhood has been transformed into a popular tourist and shopping district, which obviously affects its atmosphere, there are still reminiscences of the historical Marais: the Marais of the Templars, the Marais of the aristocracy, the poor Marais, the Marais of the Jews, the Marais of the LGBTQ+ communities, the Marais of the craftsmen, the Marais of the gallerists and the Marais of the museums. These different varieties of Marais are of course not successive or mutually exclusive. They can often be traced back to particular blocks within the Marais, such as the Jewish Marais, which was called *le Pletzl* – "the little place" in Yiddish – and centred around Rue Roisiers. Now, falafel stands jostle with designer boutiques on this street, although a synagogue and several Jewish delicatessens and bread shops still fight for space. But to this day, there are both invisible and more tangible thresholds for the casual visitor. History takes material form in buildings, streets, squares and gardens. Of course, there is a variety of architectural styles in other parts of Paris too, not just in the Marais; the variation between Gothic, Baroque, Neoclassicism and Modernism is still a distinctive feature of Paris, despite the architectural uniformity that resulted from Haussmann's rapid remodelling of Paris: more than 100,000 new buildings between 1852 and 1870.[25] Variety is one of the characteristics of the organically grown metropolis, and Naples, which was probably an inspiration to Benjamin for the writing of the work of passage, is a city where the line between old and new is less clearly drawn than in Paris (with some exceptions).[26] But what characterises the Marais is not necessarily the variety as such, but *this* very variety. Hazan's observation that crossing Boulevard Beaumarchais and leaving the Marais for the Bastille also makes one aware of changing neighbourhoods, seems correct even to the casual

25 Kirkland 2013, p. 140.
26 See Ponzi 2014, pp. 101–120.

visitor. The change of neighbourhood is experienced as a change of atmosphere.

It could be argued that the Marais is special precisely because the neighbourhood was not affected by Baron Haussmann's regulation. But the Marais is by no means unique in terms of a difference in atmosphere compared to other neighbourhoods, although it could be argued that Paris, like many other big cities, has undergone further transformations even after Haussmann.[27] Shops, cafés and restaurants have become chains spread across the city, social mobility between neighbourhoods has increased, and Paris has ever more become a global city, a given tourist destination for people around the world. This would mean that the possibility of experiencing a change of atmosphere by crossing the threshold into another neighbourhood would disappear, or simply that the threshold would be lowered, so to speak, until it was no longer possible to experience it at all.

Nonetheless, other atmospherically distinctive neighbourhood changes in Paris are discernible. In my own regular walks between the Latin Quarter and Saint-Germain-des-Près across the Odéon, the Faculty of Medicine and the statue of Danton in Place Henri-Mondor symbolise such a transition, as Hazan points out.[28] He talks about the "psychogéographie de la limite", the psychogeography of the border, where I talk about thresholds, but although "limite" is perhaps equally well translated as "border", we see from the choice of words that it is the kind of border that is a threshold.

These thresholds can be of very different kinds. Some are ambivalent or neutral zones (a no-man's land) rather than distinct lines, others are more vague. Often the historic-mythical neighbourhood does not coincide with the administrative boundary. It is not even certain that different neighbourhoods are comparable to each other just because they are called neighbourhoods. "Certain quarters of Paris have a character that owes most to history and

27 In addition to the above-mentioned works by Hazan, see in particular Hazan 2016, where he admits to harbouring a certain nostalgia for the Paris of his youth.

28 Hazan 2011, p. 3.

architecture, others to their economic activity, and others again to geography" – some other boundaries are emotional, Hazan says. I would say atmospheric.[29] Boundaries fluctuate and shift over time. But even if Hazan doesn't hesitate to admit that this is so, he can say that, despite all the changes, "*l'esprit du lieu*" – the spirit of the place – remains, as in the Belleville neighbourhood in the 20th arrondissement.[30] And then, of course, there is the Seine, with its islands, which divide the city into the Left Bank and the Right Bank, while linking them together by bridges. The inevitable change of a city does not mean that its thresholds disappear, but that these are also, inevitably, dynamic. They exist in time as much as in space. Nevertheless, Benjamin reminds us that it is not administrative boundaries that determine the extent of neighbourhoods: "Paris has more than twenty of them and comprises a multitude of towns and villages."[31] To learn to know a city is to learn to know its thresholds, and to know a city is to know its thresholds.

Paris is an educational example with its arrondissements and neighbourhoods and their historical significance, which means that many people other than Parisians themselves have experienced the importance of neighbourhood thresholds. But the experience of change when we cross a threshold between different neighbourhoods in the city is, of course, not exclusive to Paris. It recurs in smaller as well as more familiar neighbourhoods, both away and at home. On a visit to Tokyo, I stayed in a hotel in Shinjuku, near the railway station, not particularly aware that I was thus right on the edge of Kabukicho, modern Tokyo's most intense entertainment and pleasure district. When I turned into a certain street – I don't remember its name – the atmosphere changed drastically. At every gate, a well-dressed usher would try, insistently and with more or less vigour, to draw me into the unspeakable pleasures that the establishment could offer. The experience of crossing a street and suddenly changing the atmosphere to something more uncontrollable is a formal *locus*

29 Hazan 2011, p. 210.
30 Hazan 2011, p. 217.
31 Benjamin 1991, p. 999 (C°, 6); Benjamin 2002, p. 832.

classicus for contemporary travel writing, and there are far more dramatic examples in this genre than Tokyo – like being in the wrong neighbourhood in Baltimore or Chicago.

In my own home town of Gothenburg, I have the distinct experience that Bö, which is the name of the neighbourhood that is home to me, is separated from Gårda by a viaduct that passes under the highway E6. My neighbourhood is separated from one of its neighbours by a literal passage, where in some sense I actually descend into the underground. In other directions, the boundary is not as concrete; towards the neighbourhood of Kålltorp, the street Welandergatan forms a boundary, towards the neighbourhood Skår the street Delsjövägen, and so on. One definition of home could be neighbourhood competence: being able to explain to a casual visitor how to find their way around, but also having a sense of the rhythm and variation of the neighbourhood during the day, week and seasons, and an idea of who lives and moves in the area, what the church bells sound like and what other lights and smells you encounter. And finally, the thresholds that separate a neighbourhood or district from other neighbourhoods or districts. "To know them" – the cities, but Benjamin could just as easily have called them the neighbourhoods or districts – "means to understand those lines that, running alongside railroad crossings and across privately owned lots, within the park and along the riverbank, function as limits".[32] And this is true not only in Paris, but also in Tokyo and Gothenburg, and all other cities. Cities, you could say, are made up of a multitude of places and thresholds. To move around the city and cross its neighbourhoods is to move between many atmospheres, perhaps even a "chaotic multiplicity" of atmospheres.[33] What prevents the city, in an experience of pure chaos, from becoming unmanoeuvrable for those who inhabit it, are its thresholds that both unite and separate.

32 Benjamin 1991, p. 141 (C 3, 3); Benjamin 2002, p. 88.
33 Simmel 2020; Hasse 2015, p. 91 and *passim*.

The City Gates

The ancient city is defined by its walls and its gates. The wall determines very concretely what falls outside and what belongs to the city, although it can take different forms. It can be a real wall, a *pomerium* like the one that Romulus, according to legend, drew around the city of Rome, where no buildings were allowed to be built, or a mere furrow drawn by a plough to mark its location. The wall around the city not only has the function of defining the city's boundary and offering protection against enemies who want to conquer or rob it, but also marks that which falls within the city's jurisdiction; inside the wall the law prevails, outside another law or, in the worst case, no law at all.[34] When the plough that drew the furrow marking the city's boundary was lifted for a while, it indicated the gates that would allow traffic between inside and outside. No city can completely close itself off from the outside world. There must be a zone where the negotiation between inside and outside takes place; where the law begins or ends. "The act of entering through a gate is," as architectural historian Joseph Rykwert writes, "an act of convenant with those inside the walls through which the gate leads."[35] The gate is sacred: Janus, who looked both forwards and backwards at the same time, was the head of the guardian gods of gates; "he was the gate personified [...] the god of all beginnings and openings".[36] The gate implies a threshold, a zone of transit, where those who pass through it change status, are exposed to danger and are therefore in need of protection.

The symbolic meaning of the gate is linked to its material form. Even in cases where its symbolic meaning overshadows any functional significance, this does not mean that the material form disappears, but that it is stylised. In modern Paris, one finds a triumphal arch in the middle of the city, originally built as a memorial to the victory at Austerlitz in 1805, a border gate transformed into

34 See Mumford 1961, pp. 48 f.
35 Rykwert 1976, p. 135.
36 Rykwert 1976, p. 139.

a monument to victory. In Roman antiquity, "the returning general" was transformed into "a concquering hero", as Benjamin writes, but even now the gate transforms "whoever passes under its arch", at least by conferring a new symbolic meaning on those who pass under it.[37] Later stylised gates – like the Arc de Triomphe in Paris – symbolise royal power rather than entry and exit.

There are many examples. The Brandenburg Gate in Berlin was once part of the city walls but is now in the centre of Berlin.[38] Originally simply a city gate, it was eventually erected as a monument to peace and decorated with a classical Roman goddess of victory driving a quadriga. Function and symbolism in symbiosis. When it lost its function as a city gate, as the city's boundary was moved, it retained its symbolic significance: when Napoleon captured Berlin on 27 October 1806, he led his men through the gate. After defeating Napoleon, it was the turn of Friedrich Wilhelm III to pass through the gate on 7 August 1814. During the Cold War, the Brandenburg Gate once again became part of a wall, between East and West Berlin, now without the possibility of passing through it, which also came to be seen as symbolic. On 22 December 1989, it was reopened after reunification, when West German Chancellor Helmut Kohl met his East German counterpart Hans Modrow by walking through the gate. The gate stood as a symbol of reunification between West and East. Here, it unites two former rivals over what it is to be Germany, but at the same time it also separates the conflicted past from the (in principle) shared future. The Brandenburg Gate once again serves as a symbolic threshold and has done so throughout its existence, albeit with varying historical overtones. Even today, "something happens when we step through the gate from West to East", writes cultural historian Helge Jordheim, which can be likened to "passing through a portal from one dimension to another".[39]

Nowadays, the gate in the wall has largely lost its original meaning of enclosure and protection. Although the original city walls of Paris

---

37 Benjamin 1991, p. 139 (C 2 a, 3); Benjamin 2002, p. 87.
38 Jordheim 2023, pp. 133–144.
39 Jordheim 2023, p. 145.

were demolished in the 17th century, between 1782 and 1860 the city was surrounded by another wall whose primary purpose was to be able to tax those who passed through any of the city gates into Paris (strictly speaking, Paris has been surrounded by five different walls). This was demolished when the city boundary was moved to another wall further out, the Thiers Wall (a defence wall), and villages such as Belleville and Montmartre were incorporated into Paris. The Thiers Wall was in turn demolished between 1919 and 1929.

Another example: the Porta del Popolo, which is part of the historic Aurelian Wall built between 271–275 CE, is today no obstacle for pedestrians to enter Rome. You just have to watch out for traffic when crossing the street outside the gate, otherwise there are no obstacles. The real gates to the city in our time are the railway stations and airports. These can be difficult enough to get through at times and also have a lingering if transformed symbolic meaning: zones of transit that can not only be surrounded by specific legal regulations, but also transform our existential mode from traveller to someone who has arrived.[40] Even as a traveller from a well-off northern European country, it is not possible to imagine that the transition is entirely smooth. With other passports, the problems increase.

The real border gates of our time have quickly become separated from their symbolic gates. The gates into the Forbidden City in Beijing once marked a strict separation between inside and outside. Actually, Beijing was four cities, each with its own walls: the Outer, Inner, Imperial and Forbidden Cities.[41] The emperor did not appear in public and no unauthorised person could enter the Forbidden City. In other words, the gates regulated access in the strictest possible way. The main entrance of the Meridian Gate was reserved for the Emperor alone, with a few exceptions. After the 1949 revolution, Chairman Mao wanted to open the Forbidden City to Tiananmen Square; instead of high thresholds, the Tiananmen Gate – the gateway to the Imperial City – would now symbolise low thresholds. The gate would invite people into the square, instead of confining

40 See Sigurdson 2024c, pp. 127–136.

41 See Dovey 1999, pp. 72–82.

and enclosing them. Today, the Forbidden City is open to the public, even to foreign visitors like me, with no particular purpose other than to satisfy our curiosity. It is easy to walk through the Meridian Gate, the third of three gates after Tiananmen and Duanmen, into the city.

A small matter, but only if you ignore the rigorous security checks that surround the entire area, reminiscent of airport security checks. Admittedly, the Forbidden City is now basically open to the public. Above the Tiananmen Gate hangs the portrait of Chairman Mao, a symbol more recognised worldwide than most other portraits. For the casual visitor, the meaning is twofold: on the one hand, Mao symbolises the seizure of power over, and the opening of, the Forbidden City to the people; on the other hand, it is now Mao, or rather his successor, who regulates and supervises the entrance and exit of what used to be one of the highest thresholds in the world. The Forbidden City was then considered a heavenly city, where the emperor was the link between heaven and earth (another threshold), but it is not far-fetched to imagine Mao performing a similar symbolic function in modern China, albeit with different metaphysical overtones. The gates around the Forbidden City illustrate both the mutability and the permanence of the gates. In no way have they become existentially or politically trivial.

There are significant differences between the traditional gates of antiquity and those of today, but there are also similarities.[42] The gates of antiquity were indeed found in the city boundary at the wall and thus had a material anchorage where their function as a door went more clearly hand in hand with the gate as a symbol of passage. To enter the city, one had to pass through its gate. Today, neither symbol nor function have lost their material anchoring, but they have been separated. As a result, the gate's threshold function may have become more invisible.

Nevertheless, even today, gates are a reality to be reckoned with. Alongside the gates to the city that still exist in our time – the airport, the railway station, and in some cases the port – and alongside the

42 See Siegert 2012, pp. 151–170.

arches that are stylised gates without any purpose of hindering or delaying our progress – the Brandenburg Gate, the Arc de Triomphe, but also L'Arche de la Défense in Paris from 1989 – there are also gates today that more or less truthfully promise us entry into another way of existing. There are the gates that invite us into commercial buildings such as department stores and supermarkets, into public buildings such as courts, police stations and hospitals, into religious buildings such as churches, mosques, synagogues and temples, and into buildings that promise excitement and entertainment such as football stadiums, concert halls and amusement parks. While not all of these gateways lead to a different life in any radical sense, they are a call or invitation – the philosopher Louis Althusser would have said interpellation – to a change of existential mode: we are not exactly the same inside as outside.[43]

The desire of the department store, the seriosity of the court, the devotion of the church, the fascination of the museum and the intensity of the football stadium, if we are receptive to them, mean a modulation of our existential approach in which we, the addressees of an architectural and atmospheric request at the threshold, are treated to something that was not the case outside. Our time likes to emphasise accessibility and therefore many contemporary gates are designed as inviting. At least in principle: security checks can be rigorous and contradict the ideal of frictionlessness. At other times, it can be about exclusivity, where our entry is conditional.

The Doors of the House

Even in less public contexts, the gate in the form of a door plays a persistent role. According to van Gennep, "the door is the boundary between the foreign and domestic worlds" in an "ordinary dwelling" in the same way that the gate is the boundary "between profane and

43 Althusser 2014. I am sceptical about the deterministic features of Althusser's theory, see Sigurdson 2021, pp. 233 f.

sacred worlds in the case of a temple".[44] Therefore, in both cases, "'to cross the threshold' is to unite oneself with a new world". If the city with the temple at its centre was once an image of the cosmos, the home still bears features that recall the distinction between the sacred and the profane. Crossing the threshold of someone's home is a matter of course in few cultures; some prefer to socialise in the street to find a neutral zone. The home symbolises the existential centre of the individual's personal cosmos, even if it is a mobile home.

Gaining access to someone's home means staying there as a guest.[45] The host welcomes me on the doorstep of his house or flat. To be a guest is to adopt a mode of existence somewhere between that of the stranger and that of the person who inhabits a place. It means staying with a more or less temporary passport, which confers certain rights that the stranger does not have, but also certain obligations that the resident does not have. As a guest, I have access to some rooms, but far from all. Some rooms in the house are more intimate than others: the bathroom, the bedchamber, perhaps the cellar and the attic if there is one. I do not go there without special authorisation. My own initiatives are on the host's terms. As a child of a house, my image of the world is shaped by the rooms I inhabit; I draw, metaphorically speaking, a map of the cosmos that borrows its illusory or relative stability from the rooms of the home and that forms the basis of my future memories. As an adult, the home is a refuge; I can close the door on myself, but I can also open doors and windows to the outside world. The door therefore has a highly symbolic meaning and crossing the threshold marks a passage, whether I am going out or coming in.

In our time, there is widespread homelessness, increased migration and, for various reasons, many ways of living – facts that we cannot ignore. But these do not invalidate the fundamental distinction between home and away. As Michel Foucault writes, such distinctions that

44 van Gennep 1981, p. 29; van Gennep 1960, p. 20. For some significant contributions on doors, see Koolhaas, Trüby, Westcott and Petermann 2018, pp. 542–697; Simmel 1994, pp. 5–10; Latour 1992, pp. 225–258.

45 On accommodation, see Bachelard 1994, but also Coccia 2021.

serve to existentially structure space for us – such as the difference between private and public space, between family space and social space, between cultural space and commercial space, and between the space of leisure and the space of work – have not disappeared despite all the attempts of modernity to domesticate and formalise space: "all are still animated by a muted sanctification".[46] We have learnt from Bachelard and the phenomenologists, Foucault argues, that space is not homogeneous and empty, but heterogeneous and imbued with qualities and that make us associate certain affects with it. Even the distinction between home and away therefore remains valid. In a way, being on the move requires a home to relate to. Being a nomad simply means taking your home with you wherever you go. As a migrant worker, you might set up a temporary home. The relative impermanence of all our forms of home, whether house or tent, cancels neither the distinction between inside and outside nor the need for a mediating zone: some kind of door, a threshold and a way to cross it. Distinctions, such as between home and away, do not constitute a hierarchical separation; the "door", as the sociologist Georg Simmel notes, "represents in a more decisive manner how separating and connecting are only two sides of precisely the same act".[47] Therefore, Foucault's claim that our time is an epoch characterised by simultaneity and coexistence points to the realisation that we are indeed living in a threshold epoch.[48] But perhaps we should not exaggerate the difference between our time and the past. The threshold, as we have seen, is an ancient phenomenon.

But isn't it the other way round, as Benjamin seems to argue and as I have mentioned, time and again, that our time is poor in threshold experiences?[49] The social changes of the 20th century have meant that the concept of space has been homogenised, because a geometric understanding of space has come to dominate the life-world. In addition, the ideal of modernity has been to travel without friction, meaning that the thresholds should therefore be as

46 Foucault 1994, p. 754. See Bollnow 2010, p. 148, and Bachelard 1994.
47 Simmel 1994, p. 7.
48 Foucault 1994, p. 752.
49 Benjamin 1991, p. 617 (O 2 a, 1); Benjamin 2002, p. 494.

low as possible, if not non-existent. Our time values our capacity for independent action, our agency, highly. But the thresholds allow us to experience something that we cannot fully master and control. The lower the thresholds, therefore, the greater the scope for our own independent action. A Haussmannian regulation, so to speak, has been implemented for our entire existence: wide streets and homogeneous façades, metaphorically speaking, in every possible aspect, including doors.

In an essay by Robert Musil entitled "Doors and Portals", the author ironically claims that "[d]oors are a thing of the past". You used to be able to eavesdrop behind doors, but now everything is heard through the walls. Doors could be representative and show the social status of the owner, but now their function has been replaced by "other objects that satisfy this same purpose: travel, cars, sports, winter vacations, suites in luxury hotels".[50] They have lost both technical functionality and representational value. "The great age of doors is behind us!", Musil exclaims, and so are thresholds.[51] And in a reflection in *Minima moralia*, the philosopher Theodor Adorno strikes a similar note when he argues that the technologisation of life leads to an instrumentalisation of the interaction with things. Living experiences of our surroundings are replaced by goal-orientation and "mere operation" and what falls outside this handling, "all hesitation, deliberation, civility", is forgotten:

> Thus the ability is lost, for example, to close a door quietly and discreetly, yet firmly. Those of cars and refrigerators have to be slammed, others have the tendency to snap shut by themselves, imposing on those entering the bad manners of not looking behind them, not shielding the interior of the house which receives them.[52]

Especially the latter, the automatic doors, seem to neutralise or conceal the threshold experiences Benjamin writes about, when all emphasis is on human agency. From such a small detail as the art

50 Musil 2006, pp. 59, 61.
51 Musil 2006, p. 62.
52 Adorno 2005, p. 40.

of closing doors, Adorno develops how subjectivity is transformed under modernity.

Musil's and Adorno's comments on the changing status of doors in the first half of the 20th century are probably an ironic comment on functionalism, especially those features that emphasise transparency, frictionlessness and purposefulness.[53] Although there are other features of this style, the two authors pinpoint a trend in technological development in the second half of the 19th century and throughout the 20th century. Doors are, one might say, a technology for managing the inevitable distinction between inside and outside in every sense of the word – economic, epistemological, social, symbolic – in every culture, by being both closed and open. As the technology of the door develops in the 20th century and the classic door that we have to open and close ourselves is increasingly replaced by automatically opening doors, by swing doors and by labyrinth entrances, or doors that we never have to touch, the relationship between inside and outside is renegotiated. The automatic door in all its forms means that the threshold between inside and outside is, if not disappearing, at least being reduced. Against the discontinuity marked by the threshold, the automatic door juxtaposes the continuity of movement: with automatic doors, we not only avoid interacting with things but are also allowed to move without interruption.

"The only original door conceived by our time is the glass revolving door of the hotel and the department store" writes Musil.[54] The revolving door was invented in 1888 by the American Theophilus Van Kannel, who won a prize for his invention.[55] Paradoxically, one of its original purposes was to keep the weather out, as in a sense it is always closed. The freedom of passage provided by the revolving door also meant increased regulation of the indoor climate, further separating it from the outdoors. "Always Closed" was used early on in its marketing, but eventually also in parallel with the motto "Always Open", as an illustration of this paradox. As a central

53 See von Arburg 2011, pp. 319–354 and also Jonsson 2000, pp. 60–96.
54 Musil 2006, p. 62.
55 See Buzard 2001, pp. 559–581.

symbol of modernity in the 20th century – not only through its association with skyscrapers, large hotels and commercialism, but also early on as a recurring prop for the emerging film industry – the revolving door exhibits the paradoxical aspects of modernity.

Against the ideal of the frictionless passage stands the concrete experience of the revolving door in a kind of parodic version of the dialectic of enlightenment. Neither of the two revolving doors I use with any great regularity – in the workplace and in the supermarket – works very effectively. Both are automatic revolving doors, which means that instead of pushing when I am about to pass them, the point is that I should absolutely not push. They should turn by by themselves. However, there are several reasons why they stop: someone else or I accidentally push anyway, there are too many people in the door at the same time (the gaps between its panels are individualistically conceived) or the mechanics simply don't work. Instead of flowing through, the passage comes to a standstill, you have to push much harder than a normal door and, in fact, the choice of the manually operated door alongside the revolving door would have been more efficient. The invention of the revolving door gave rise to a number of legal dilemmas about who was responsible for the accidents that followed its deployment. So it is not as easy to get through as was intended.

This example may seem trivial, but it points to a more general experience of the supposed effectiveness of many of the technological solutions that are supposed to lower thresholds and create an increasingly frictionless existence.[56] In addition to the endemic vulnerability of us corporeal beings that contradicts the promise of all such technologies for a more frictionless existence, increased security controls and surveillance are part of this. The revolving door is supposed to let me into the workplace, but I still have to show my access card and punch in my personal code. The building that houses my (now former) workplace is a modern building with doors that open automatically as well as some revolving doors. Despite this, it has less accessibility than any previous version of the

56 See Ellul 1964.

same building, as security now requires an intricate system of locked doors to prevent unauthorised entry. In other words, an architectural emphasis on accessibility and transparency – large windows that cannot be opened – has been combined with greater inaccessibility and enclosure. The revolving door does not abolish the threshold as such. Instead, it is a machine of automated transport between inside and outside that epitomises Adorno's description of the forgotten art of walking through a door both silently and firmly.

When doors cease to be tools that we can use and become machines that regulate our entry and exit, thresholds have not disappeared at all. They have just changed shape and become less visible. There is still a legitimate need for doors and thresholds to demarcate between home and away or between me and you. Without these, we would live in what anthropologists call a permanent liminality or simply in chaos. Therefore, there is a need for threshold knowledge, the competence to cross thresholds. We should, like Benjamin, understand the thresholds dialectically: even the thresholds that should be lower remind us of the need for a certain height in others. Not all thresholds are good, but neither is every endeavour to lower thresholds by definition liberating or beneficial. My point is that they exist and influence how we relate to space. To cross the threshold is to change our approach.

Threshold Oblivion

There are still thresholds between cities, neighbourhoods, buildings and homes in our time. But is it really such a big deal? Is crossing a concrete, material threshold really always a threshold experience? Isn't it often simply a matter of exchanging one room for another, without any major consequences for our existential stance?

Bridges and doors are united, according to Georg Simmel, in the concrete materialisation of one of humankind's greatest achievements, namely to connect two places by building a road between them. Both bridges and doors not only bring about such a union, but also visualise it. They differ, however, in that bridges are more concerned with bringing about and visualising the connection between two places, while doors make it clear that any connection also presupposes a separation. The door cancels the distinction between inside and outside by both preserving and annihilating it. Bridges and doors thus become emblematic of two different approaches to passage:

> Whereas the bridge, as the line stretched between two points, prescribes unconditional security and direction, life flows forth out of the door from the limitation of isolated separate existence into the limitlessness of all possible directions.[57]

The bridges connect "finite with finite", while the doors represent a junction of the connection between finite and infinite in a physical and metaphysical sense. In other words, the bridges emphasise continuity, while the doors symbolise a more dialectical relationship between continuity and discontinuity. It also doesn't matter which way we cross a bridge, but at the door it matters whether we are entering or leaving. We can close a door to prevent passage, but we can also open it to pass through. We can close the door around us because we want to be left alone, thus separating us from the outside world, or we can open the door to join the outside world. The door, as Simmel points out, combines limitation with freedom. Insofar as we humans are threshold beings who combine limitation with freedom, the door is a more pregnant symbol of human nature.

The bridges of Paris may connect two banks, the left and the right, and thus act as a threshold between the different atmospheres of the various neighbourhoods. But in all honesty, I have never experienced a major change of atmosphere when travelling by car, bus, tram or on foot across the bridges that connect the city Gothenburg to the island Hisingsön. The experience is more, as Simmel describes it, of a line between two points that prescribes a certain direction without any discontinuity. On the other hand, of course, there are also those bridges that are really a kind of threshold, such as the pedestrian bridges over the Tisza River that connect Ukraine with Romania

57 Simmel 1994, p. 8.

and facilitate traffic of various kinds between the countries, even during wartime. But aren't most bridges more like the bridges linking Gothenburg to Hisingen, that we don't think much about our passage over them? Are not our passages in general, especially the more mundane ones, characterised by the continuity of the bridges rather than the discontinuity of the doors and thresholds?

We can turn this around and ask ourselves whether all the concrete material thresholds I have described in this chapter, from the thresholds between neighbourhoods to the thresholds of gates and doors into and within our homes, are not in fact quite low? Crossing them would then be more like walking across a bridge than going through a door. Is it not we ourselves who in some sense choose to consider them as thresholds, rather than their concrete materiality causing a threshold experience that in some minor or major sense changes our lives?

In one of the fragments from *The Arcade Project*, which I have already quoted in part, Walter Benjamin writes

> These gateways – the entrances to the arcade – are thresholds. No stone step serves to mark them. But this marking is accomplished by the expectant posture of the handful of people. Tightly measured paces reflect the fact, altogether unknowingly, that a decision lies ahead.[58]

Here Benjamin is talking about thresholds that are not marked materially but registered bodily, through the pedestrian's sparingly measured steps before the decision to cross the threshold or not. Being faced with this decision is, as Benjamin puts it, not the result of an act that we consciously perform but something that happens spontaneously; subliminally if you will, or in other words "below the threshold" of our consciousness. Benjamin here conveys an essential insight into how the thresholds I have highlighted in this chapter usually function: at least in neighbourhoods, houses and dwellings that are familiar to us, they do not constitute obstacles because we have acquired a locally conditioned threshold competence through habit. That there are threshold experiences of more dramatic kinds

58 Benjamin 1991, p. 142 (C 3, 6); Benjamin 2002, p. 89.

even in our time is the conclusion I would like to draw from the chapters that have dealt with passages of various kinds, as well as revolutions, conversions, therapies and limit situations. But if Benjamin is right, the concrete thresholds have not disappeared from our everyday lives either. We may not notice them to the same extent as the more dramatic ones, but we register them physically.

Habit is a powerful force in the invisibility of the more mundane thresholds, but whatever their visibility, it is first and foremost in the bodily experience that we register them. Although the height of the city's thresholds has an inescapable cultural and social dimension, not just a physical-material one, and access to a place is regulated by our own resources of economic, cultural and social capital, these dimensions are intertwined with what we register with our bodies. Getting to know a city, a neighbourhood, a building or a home is getting to know where to go, at what time and with whom or, in other words, where the thresholds are. To get to know any of these places is to acquire a place-specific habitus.[59] Such a habit can be described as an acquired disposition to experience a place as a familiar place where there is thus a continuity between my actions and the place. But at the same time, habit is not opposed to experiencing discontinuities within the framework of the familiar place. The skill of the accomplished tennis player does not consist in being able to anticipate every aspect of the game, but in having a repertoire of useful and varied ways of responding to the situations that arise in the course of the game. The skilful tennis player does not even have to consciously think about how to respond to the opponent's strokes, but, as in the quote by Benjamin above, the return the player faces is reflected in the steps. The thresholds represent a break in the continuity that the habit strives for, but the habit also provides tools to deal with such discontinuities.[60] Habit does not necessarily grind down any thresholds, but it inserts them into our repertoire of responses to the places we inhabit. We become skilful in dealing

59 See Bourdieu 2002, pp. 87–95. However, my discussion on habit is mainly influenced by Ravaisson 2008.

60 See Hasse 2020.

with the thresholds we encounter, so skilful that we risk forgetting their existence. If habit becomes routine, then thresholds tend to be forgotten. But habit, as I said, is not the antithesis of change: habit comes from change and also generates change. A human being is a threshold existence whose essence is found neither in habit nor in change, but in the indissoluble tension between the two. Simmel is right that we are, as human beings, more like the door than the bridge.

There is a tendency, at least among the thresholds we are more familiar with, for them to disappear from our consciousness. Combined with a specific forgetfulness in modern society's relationship to place and history, and a drive to lower each threshold to maximise our agency, their relative invisibility risks increasing.[61] As a result, they will eventually disappear from our imaginary. It is such a process that Benjamin describes. The disappearance of thresholds as a central symbol in the imaginary world in which we understand our lives is serious enough, but the remedy is then more about recalling them from oblivion than rebuilding them. Whether thresholds should be lowered or raised is a question that must be asked specifically and cannot be answered in general terms.

*

Why do vampires need to be invited across thresholds? Like us ordinary people, vampires are threshold creatures, but more than the rest of us, they themselves are aware of this. The invitation is more about us than about them. It is a reminder that every crossing of the threshold involves a change in our existence that is filled with both danger and possibility. It's not just about swapping one room for another. We never really know what awaits us when we cross the threshold. We don't know who we will meet, even when it comes to ourselves.

61 See Connerton 2009.

Part 2

Places

Place Saint-Sulpice

Date: 29 May 2023 (Whit Monday). Time: 9:12 a.m. Place: Café de la Mairie at Place Saint-Sulpice in Paris.

The sun hasn't yet had time to warm up the air, it's a bit windy, so I've chosen their glazed veranda. Here I have a view of the place from behind a round Formica table with a brass frame and iron base. The café staff are classically dressed, with black trousers and shoes, white shirts and black bow ties; someone also wears a black waistcoat and black apron. Garbage collectors in green overalls and yellow vests come by for a quick takeaway coffee in a paper cup. The green bin lorry then moves on through the area. Everyone is greeted as they arrive, including me: "Bonjour monsieur, bonjour madame." I order a double espresso. "Manger?", "Non, merci." I then change my mind and eat a croissant. From the conversations I hear that most guests, at least at this time of day, are regulars. Some stay longer, like me. Some read, talk on the phone or chat to each other. Others stay just a few minutes, for a coffee at the counter. "Bonne journée." I wonder if it's particularly quiet today because it's a public holiday?

The floor of the Café de la Mairie is a worn mosaic floor with a geometric pattern. The toilet smells of strong disinfectant with a hint of chlorine; for me, the smell brings back memories of visits to my parents when they lived for a few years in the south of France. They must have used a similar cleaning agent. Outside the café, a double row of chestnut trees stands like a colonnade. Through their foliage, light plays down on the street. A moped rumbles past. Don't they sound a little different from home? A lorry covered with a large, green Heineken sign passes by. The church bell strikes ten.

Otherwise, I mostly hear the slam of metal pipes falling to the ground. A remodelling is underway in the square. The famous fountain with statues of the four Christian orators built by Joachim Visconti in the 1840s is surrounded by fences and small tents, which means that I can never get close on this particular visit. (The following year the tents will be removed and the view will be clear when I revisit the site.) The best view of the fountain comes when you just pass out of the church through the gate facing the square. Église Saint-Sulpice – one of the largest churches in Paris – rises imposingly into the sky. It was here that both the Marquis de Sade and Charles Baudelaire were baptised and where Victor Hugo married Adèle Foucher. When I looked in before visiting the café, I saw long rows of chairs with backs that looked almost abstract in the morning sun. I am now sitting on a wicker chair, leaning my left arm against the neighbouring chair while I write in my notebook.

I'm here for a reason, to deepen my awareness of this place.

The choice of this particular place is not random. The Café de la Mairie is one of the three cafés on Place Saint-Sulpice where Georges Perec writes his *An Attempt at Exhausting a Place in Paris*. On 18, 19 and 20 October 1974, Perec sits in the same café as I do now, as well as in two other cafés that no longer exist and on a bench in the square, and takes notes on what is going on during his visit. Perec, however, is not interested in all the historical monuments that make up the place – the church and the fountain – or the famous people who have left their mark here. Instead of the "extraordinary", he tries to capture the "infra-ordinary", that which is "below" the ordinary.[1] As he puts it in his preface, there are so many attempts

1 See Perec 1997, p. 210. This chapter, "Approaches to what?" was originally published in 1973 and can be seen as a kind of programme statement for Perec 2020. In fact, the notion of the "infra-ordinary" in Perec is not the subject of conceptual development, but is passed over in a list of what should be depicted in an attempt to approach everyday life, but which rarely appears in the depiction of it: "the banal, the quotidian, the obvious, the common, the ordinary, the infra-ordinary, the background noise, the habitual". However, the infra-ordinary has proved productive in analysing everyday life precisely as a reminder of the forgotten. See,

to document what goes beyond the ordinary, that he himself wants to devote himself to what remains: "what happens when nothing happens other than the weather, people, cars and clouds".[2]

Whenever Perec's little book is mentioned, it is customary to emphasise melancholy in that such an attempt is doomed to failure. It is not possible to exhaust everything that happens. The limitations of all description, together with the flight of time, make the endeavour impossible. That may be true, but doesn't the idea of the infra-ordinary in any case contain an important insight into what makes a place a place? In his first notes, Perec describes what is "strictly visible" (letters, numbers, asphalt, trees, pigeons, vehicles, people, lettuce, etc.), the direction of different vehicles, colours, and then goes on to expand these investigations. These enumerations serve to make us aware of how we relate to all those "tens, hundreds of simultaneous actions, micro-events", all those infra-ordinary events that create the place we recognise and that somehow make us recognise it without us having to recall it in our conscious memory.[3] If it is a place we recognise and feel at home in, as Place Saint-Sulpice undoubtedly was for Perec, then it belongs rather to the stream of perception that constitutes our life-world even when we are not thinking about it.[4]

During the three days that Perec takes notes, he sees acquaintances who pass by, sometimes several times. Occasionally he also greets them: Jean-Paul Aron, Jean Duvignaud (perhaps), Geneviève Serreau, Michel Martens (with a geranium umbrella).[5] Even Paul Virilio passes by on his way to the Bonaparte cinema nearby.[6] Virilio is the philosopher who, together with Perec and Duvignaud, edited the journal *Cause commune*, where the text I am discussing here was first published, a journal that explicitly wanted to examine

---

in addition to my own attempt in this book, how Frykholm 2025 uses the concept in an architectural-theoretical exploration of his apartment, and also, for a more philosophical exploration, Hasse 2017, pp. 21–33.

2 Perec 2010, p. 3.

3 Perec 2010, p. 10.

4 Cf. Cosnard 2022, pp. 100 f.

5 Perec 2010, pp. 18, 22, 29, 42, 45.

6 Perec 2010, p. 33.

the everyday, the banal and the infra-ordinary. It was Virilio who, somewhat earlier, had asked Perec to write *Species of Spaces*, a novel published the same year Perec wrote *An Attempt*.[7] The mention of Virilio is not only a literary nod to his friend and colleague, but also brings the list of infra-ordinary events to life. The sixth arrondissement is a familiar neighbourhood for Perec. His friends also live here. He himself is recognised here.

Perhaps Place Saint-Sulpice is a place where Perec feels at home in some sense. That is not the case for me. I have probably passed by the square on some previous visit to Paris without really thinking much about it. Now I actively seek it out, having read Perec's book. But, in fact, I am a tourist on a temporary visit. Although for a few days every morning I walk to the Café de la Mairie from the Hotel Atmospheres on Rue des Écoles, past the Collège de France where the philosopher Maurice Merleau-Ponty had a chair, that doesn't exactly make me a regular. No one recognises me here; I am anything but at home. For me, this is an interesting place, partly because Perec sat here and wrote, and partly because it is a famous square in Paris. My hope is that, with the help of Perec, I can learn something about what makes a place a place by both visiting the place and letting myself be guided by the author to, in the words of the philosopher Iris Murdoch, "see more".[8]

But I am not the first to make such a return visit. From 14 to 16 January 2020, the Danish poet Morten Søndergaard visited Place Saint-Sulpice and the Café de la Mairie. This is where he is writing his book *Forsøg på at aflytte et sted i Paris* – "an attempt to listen in on a place in Paris". Søndergaard has noticed how *An Attempt at Exhausting a Place in Paris* is dominated by visual impressions and therefore decides to go to the same place, but to eavesdrop on the place instead; to *listen* as closely as possible to what is happening there, not just to look.

Like Perec, Søndergaard spends three days at the site. Like Perec, he is preoccupied by the passing buses (which, incidentally, stop just

7 See Walker 2001, pp. 15–18; Perec 1997.
8 Murdoch 2014, p. 31.

outside the Café de la Mairie). Like Perec, the attention wears him out. Like Perec, he notices the pigeons. For the inattentive, it is easy to miss that both are also ordering the same dish, or at least say they are: some sausages, with a glass of Bourgueil wine.[9] Søndergaard's change of sensory organ, so to speak – from eyes to ears – makes us, as readers of both these books, more aware that the infra-ordinary is not limited to a single sense, or even two. Taste (the coffee, the sausages, the wine), smell (the disinfectant), touch (the buttocks and loin against the woven seat and back of the chair, the wind and the sun against the skin), as well as the very fact of existence (sitting on the glazed terrace rather than inside the café or outside at the tables in the street) influence my experience of the place and the café. Even the weather, as Perec points out, has an impact. My first visit here is in early summer, not autumn or winter – days filled with sun, without clouds or rain. On a later occasion in autumn, it is pouring rain. So it's not just about "seeing more", but also about listening more, smelling more and so on.

Experiencing a place like Saint-Sulpice is essentially a synaesthetic experience that is sensory-bodily before it is cognitive. In a certain, definite sense, both Perec's and Søndergaard's texts come too late; they are secondary accounts of primary experiences in that they make us aware of the role of the infra-ordinary in the creation of a sense of place. Through a thousand variations they can say something important about the atmosphere of the place and thus convey something of it, but they relate to the atmosphere of the place and do not replace it.

What is also clear to both Perec and Søndergaard is that the place itself exerts some kind of initiative. Both their texts are very concrete in terms of time and space and specific in their description of infra-ordinary events, but they are also to some extent free of autobiographical features. Although their descriptions of all the events of the place are dependent on their being there and being able to sense them, what we find out is precisely about the place. This is also the key to why I, as a casual visitor, am not completely removed

9 Søndergaard 2020, p. 65; Perec 2010, pp. 18, 33.

from the context of the place. True, I don't see anyone I know (a passing scooterist reminds me of a good friend who is a philosopher, but I know he is in Sweden), but all the "micro-events" of the place itself that meet my senses invite and prompt me in ways that could have eventually turned into a lasting relationship, if I moved to Paris. And I too, with my presence and eventually my writing, become a – if ever so modest – "micro-event" in the history of the place. Søren Kierkegaard was perhaps right that it is never really possible to return to a place you have previously visited, let alone a place you have previously mostly read about.[10] But is it this relative impossibility that also creates the possibility of visiting a place at all? If a place consists of a non-identical repetition of what Perec spoke of as "tens, hundreds of simultaneous acts, micro-events", all the infra-ordinary events, isn't it precisely this repetition that makes the place open to new visits?

In Place

What makes a place a place?

To be "in place", to "take place", to "place" describes our existence in relation to a point from which we orientate ourselves at a given moment. Orientation is rarely exclusively geographical, sometimes not at all. Announcing that we are "in place" can be a message that we have arrived and are now ready for what lies ahead. The place in question has an existential or functional meaning and cannot be reduced to a location whose coordinates would exhaust its meaning – although if it is a concrete place, it can of course also be marked on a map. Because we humans are corporeal, affective beings, we can never escape our place-boundness in this double sense, as both existential and geographical.

Mostly, place-awareness is about experiences that are difficult to articulate. Primarily because they are so mundane that they are simply beyond our attention, but also because they are so complex

10 See Kierkegaard 1997.

that analysing and articulating them requires a combination of observational and expressive skills that hardly comes naturally. "I am home now" – a statement like this presupposes and contains so much more than information about which geographical location I am in at a given time. It may be the beginning or end of a story about an adventure. Our spontaneously given knowledge of places is practical, rather than theoretical, because our relationship to the world is not only cognitive, but first and foremost sensual, embodied, material.

For this reason, a theoretical investigation of place as a concept and phenomenon assumes less the character of an elucidation of something previously unknown and more of a rehearsal of something we all know but presuppose rather than formulate. Such an enquiry cannot therefore be likened to opening a door to a previously unknown space. But this does not mean that places are inaccessible to the creative power of language. One of the tasks of the art critic is to deepen our attention to the work of art, to, as Murdoch puts it, "see more" in it than we have done before. This presupposes, however, that we are facing the same work as the art critic and that we are to some extent familiar with the critic's conceptual world. If both of these conditions are met, we can return to works that we may have studied before, but now with deeper insight. Our ability to pay attention to the details that characterise the work has increased and we can therefore get to know it better. Our experience of the work is richer thanks to the art critic's exposition. This is also the case with some literature, as well as with philosophy in the broad sense. They help us to "see more" – in the sense of a deepening of attention – of what it is that makes a place a place. In what follows, I have therefore drawn on writers, artists, architects, philosophers, geographers and others to sharpen my attention to concrete places, and to deepen my insight into what characterises place as a phenomenon and a concept. I move between the concrete and the abstract, between site visits to places familiar and less familiar to me and the reflective discussion of the concept of place, and back again: the return visit as a philosophical method. I myself have become somewhat familiar with the places I discuss here by repeatedly visiting them.

Maurice Merleau-Ponty, in *The Phenomenology of Perception*, is concerned with a similar distinction between attention and exposition, and recounts, among other things, the first time he stepped out of the train station in Paris. Encountering its streets was like encountering the first words of a stranger, "manifestations of a still ambiguous essence, but one already unlike any other".[11] Our natural experience of the city, for example, does not consist of a number of separate perceptions of objects that we can compare with others from these separate perceptions, but instead consists of "a flow of experiences".[12] Admittedly, it is always possible to dissolve this stream into a number of comparable perceptions, as when one's eyes are caught by a brick in the wall of the Tuileries in the Place de la Concorde and we lose ourselves in "this yellowish, gritty surface".[13] But every such perception is secondary. What we perceive first is an urban figure in the same way that we perceive a human figure:

> Just as a person gives evidence of the same emotional essence in his gestures with his hands, in his way of walking and in the sound of his voice, each express perception occurring in my journey through Paris – the cafés, people's faces, the poplars along the quays, the bends of the Seine – stands out against the city's whole being, and merely confirms that there is a certain style or a certain significance which Paris possesses.[14]

Like the person we perceive, we experience a latent meaning of the city that is bestowed on us through these primary experiences and which is the very basis of everything we might want to say about Paris, so long as we don't rely only on second-hand representations.

What these reflections from Merleau-Ponty can help us with at this initial stage is a first indication of direction for the investigation of what it is we experience when we experience a place: the concrete encounter with a place, not primarily as a cognitive knowledge but as an atmospheric, "holistic" experience. Sensing the atmosphere

11 Merleau-Ponty 2005, p. 328.
12 Merleau-Ponty 2005, p. 327.
13 Merleau-Ponty 2005, p. 342.
14 Merleau-Ponty 2005, pp. 327 f.

of a place is more akin to spontaneously pulling the covers over oneself while still asleep because one is cold, as opposed to waking up, analysing the problem and doing something about it. The latter may of course be appropriate, but it is a secondary process, not an atmospheric experience.

In addition to some concrete places that are directly associated with my reading and writing – two offices – and which are more generalisable than generally accessible, I have here somewhat arbitrarily selected three places that help me think about place as a concept and as a phenomenon: the classic square Place Saint-Sulpice in Paris, the district of La Défense, also in Paris, and Kungsportsplatsen in Gothenburg. The two places in Paris are less familiar to me, but I have visited them regularly, precisely to deepen my knowledge of some places that are not spontaneously part of my everyday habitat. Kungsportsplatsen, on the other hand, is part of my everyday repertoire, as it forms a kind of hub in my home town of Gothenburg.

Together, these places – without being particularly spectacular in nature – shed light on different aspects of place as a phenomenon and as a concept. As Murdoch has pointed out, our words depend on our attention, just as our attention is affected by words. The revisiting of these places, not only concretely but also in the reflection that follows, can therefore serve to deepen the understanding of place as such, and not simply as a deepening of the understanding of these concrete places that the reader may or may not have visited at some point. Like Merleau-Ponty on his first visit – he later came to live and work in Paris not very far from the place with which I have begun this chapter, Place Saint-Sulpice – the first impression is that of "a still ambiguous essence, but one already unlike any other".

Articulating the Experience of a Place

Merleau-Ponty's "flow of experiences" is akin to what Perec calls the "infra-ordinary". To "be familiar" with a place, or even to "feel at home" there, constitutes an act of synaesthetic perception that is not directly dependent on an active act of consciousness. On the contrary, it is usually given as a background to everything we are actively engaged in, whether it is drinking coffee or taking notes for an upcoming writing project. Even for the casual visitor to the Café de la Mairie, the visit provides an atmosphere that we might call "Parisian café" for the sake of simplicity, but which for Perec must have meant a sense of belonging. If someone were to ask me what characterises the atmosphere of a typical "Parisian café", I would be in trouble, because it would make great demands on my powers of articulation. Probably, I would take my cue from Perec and other writers who have depicted some aspect of café life. Nevertheless, like Perec, I would fail in my endeavour, not only because the flow of perception runs below my consciousness, but mainly because it is synaesthetic and potentially inexhaustible.

But at best, the failure would be a successful failure, so to speak. Compared to the sensation of a place, any attempt to describe a place will suffer from an articulation deficit. This deficit has at least two aspects: the synaesthetic character of our perception which will inevitably be dissipated by the analytical character of language, and the in principle inexhaustible richness of the objects of perception.[15] This in no way means that language is our enemy: what it holds up to the light of consciousness serves to articulate and deepen our sense to some extent. Language co-operates with perception, even if it does not exhaust it.

Thanks to Perec and Søndergaard, I know something more about what it means to sit in the Café de la Mairie and look out onto the Place Saint-Sulpice, and because of this experience I also know something more about what makes a place a place.

Since I am a place-bound human being by virtue of my physicality, the investigation of place as a concept and phenomenon will always be based on the experience of being *in medias res*. Wherever I am, whether at home or away, the concrete place will be the condition

15 See Nietzsche 1998, p. 149: "all concepts in which a whole process is summarized in signs escape definition; only that which is without history can be defined".

for me to conduct such an enquiry. In a famous speech in the House of Commons on 28 October 1942, Winston Churchill uttered the oft-quoted phrase that “we shape our buildings and afterwards our buildings shape us”.[16] He is only half right, because it is hardly the same “we” who shape and who are shaped. Someone else – in history or in the present – has formed the buildings and places we inhabit or find ourselves in, and now they are forming us who are their current and actual inhabitants and visitors. Those who once built the concrete places we now occupy were in turn in other or earlier places that formed them without their immediate involvement. The act of living or being in a house, a place, or a city is an expression of a process of tradition in the literal sense of the word, namely handing over, whether it is a handover between generations or in the present. Annie Ernaux writes in *The Years* – sometimes described as a collective autobiography – about how this process of tradition is transmitted through habits and body movements that depend on place: “Memory was transmitted not only through the stories but through the ways of walking, sitting, talking, laughing, eating, hailing someone, grabbing hold of objects. It passed body to body, over the years, from the remotest countrysides of France and other parts of Europe”.[17] We are always already shaped by places as we shape other places which in turn shape us.

For this reason, the chapters that follow are concerned with articulating, in a phenomenological tradition, different aspects of the concept of place by moving around the experience of concrete places in a constellation of perspectives, both synchronic and diachronic. In the next chapter I will explore the relationship between the concepts of place, locality and space, so as to clarify what we mean when we talk about place, and to ask what it means to have a place and in what sense places are something we share with each other, not only in the present but also as memory. The following chapter will be devoted to the question of whether place is created or found, that is, to what extent it is we ourselves who actively create our places and to what

16 Churchill 2025.
17 Ernaux 2017, p. 26.

extent it is the places that, through atmosphere and perhaps even a spirit of place, create us. The adjacent and penultimate chapter discusses the alleged placelessness of modern human beings, i.e. whether in our time we lack an existential foothold in a concrete place from which we are able to orientate ourselves. In the final chapter, the focus is instead on topophilia, the love of a particular place, often the place we call home, and the extent to which this particular sense of place constrains us or is a prerequisite for our ability to orient ourselves.

All chapters in this second part of the book are informed not only by the question of what makes a place a place, but also by how place as place connects us to the familiar as well as the unfamiliar. Getting to know a place – not primarily the concept of place but a concrete place – means both getting to know it better and recognizing the unknown in the midst of the known. As we know, we are not always at home, whatever that word means, and even when we are at home we always relate to what is away. We are always placed somewhere, but being in place is not just standing still. It is only the dynamic repetition that makes a place a place – for us.

Office/Kungsportsplatsen

As I write these sentences, I am at latitude 57° 41' 40.6" N and longitude 11° 59' 04.6" E. Or, in other words, in an office at the Faculty of Humanities at the University of Gothenburg. Latitude and longitude are invaluable for determining the location of a place and for finding it. But latitude and longitude say nothing about what the place looks like and what it means to us. In this case, is it an office with lots of light or is it dark? Am I surrounded by books or is it just me on a chair and my computer on a desk? Am I surrounded by colleagues in neighbouring offices or am I sitting alone in a corridor? Does the department that provides me with the office also provide a good intellectual environment or is it more like a research hotel? In and through all these questions, but perhaps more intangibly, there is also the question of whether the office has taken on a personal character, whether I have *settled* into the office: am I surrounded by *my* books and perhaps also *my* things? Do I *enjoy* it? The writing process is highly personal, but it is hardly unusual for people who devote a large part of their time to writing to more or less habitually seek out certain places that are conducive to it in various ways: an office in a workplace, a particular place in a library, a dedicated writing space in the home.

A geographic location can be defined by coordinates that precisely indicate its latitude and longitude. However, the term "place" can have at least two different meanings in everyday language. A place does not have to be just any place that can be easily exchanged for another place. It can also belong, in a way that remains to be explored, to what makes me who I am.

In this chapter I want to get at a particular experience, an experience of place, which, like other experiences, is an intertwining of sensations and language. In other words, I am writing about that which is sometimes called philosophical topography or topology, the experience of place, in order to understand how its different components are interrelated. I am writing about what makes a place a place, especially in this latter, more existentially significant sense.[1] My office is such an existentially significant place, at least for me. But to make it clear that the act of turning a place into a place is not a one-person endeavour but a social act, in this chapter I will move out from the office to Kungsportsplatsen in Gothenburg. Kungsportsplatsen also takes its place in my personal psychogeography, but precisely by virtue of being a shared and historical place. Place and person are connected, and the most important places are those we share together, those that in some sense precede us all.

Place, Locality, and Space

An existentially significant place cannot be reduced to a geographical location, although it often presupposes one. But its meaning can be characterised in many different ways. So I have become accustomed to writing in my office and have therefore set myself up there in a way that allows the space to support my writing. But this doesn't mean, for example, that I've organised my office in such a way that it has, as far as possible, a kind of homely feel – an atmosphere of home. Sure, there are personal touches in the office where I'm currently writing – my books on shelves and in stacks on the tables, some etchings by Piranesi on the walls, a bronze Neapolitan *cornetto*, two coffee mugs – but that doesn't make the office *mine* in the same sense as my *home*. Other environments where I have enjoyed writing are various university libraries around the world, such as the Bodleian Library in Oxford or the library of the Swedish Institute in Rome. These libraries have

1 See Malpas 2018, p. 15.

no sense of home, but rather an atmosphere of historical scholarship and diligence in relation to the readily available reference books that adorn the shelves of the reading room and the (at least in Oxford) almost endless book stacks. The significance of *the workplace* is distinctly different from the existential significance of *the home*. My point so far is very simple: different places can be atmospherically charged in many different ways.

It may also be worth stating, at the outset, that the existential significance of places is not usually dependent on my having made them so. I once stood before the Wailing Wall in Jerusalem and overwhelmingly experienced its significance, without having any personal connection to the place. The meaning of places is not only something we create by settling there, but also something in some sense already given that we appropriate in different ways. As I said, making a place a place is not a one-person job, even in the case of the home, but a social act. Making a place a place is therefore not the same as making a place one's own. The place precedes, in a certain sense, the individual initiative and meets us both as a given, to some extent, and as a request.

The existentially significant place establishes a "here" that does not only involve geographical locality. Faced with the Wailing Wall, when I realised that I was really there, that is, "here", this was not primarily an experience of being at latitude 31° 46' 36" N and longitude 35° 14' 3" E, but of being in a delimited place where something important is present.

However, the "here" of place can also be compared and possibly contrasted with another spatial experience, namely the experience of what we might call "space" or even "room". There are differences between languages: the Swedish terms *plats* (place) and *rum* or *utrymme* (space) do not capture as clearly the contrast made, for example, in English between *place* and *space,* in French between *lieu* and *espace*. Although none of these terms are semantically equivalent across languages, historically they go back to the distinction between the Greek *topos* and *chora* or *stadion*, as well as the Latin *locus* and *spatium*. German uses the same word as Swedish, namely *Raum*, but here it is possible to distinguish between *der Raum* and *das Zimmer*. *Rum* in Swedish usually denotes the latter, i.e. a delimited room in a building, for example an office in a workplace or a bedroom in a home, but such a room can of course be just a place. However, the concept of space that I am trying to compare with the concept of place denotes something more abstract and, above all, something that, unlike a particular room in a building, is not defined as bounded, but rather the opposite: a space that is open, a space of possible actions.

The experience of place is distinctly different from the experience of space. Human geographer Yi-Fu Tuan, who has devoted an entire book to this distinction, describes it this way: "Place is security, space is freedom: we are attached to the one and long for the other."[2] According to Tuan, both are essential aspects of human existence, security as well as freedom, and human life is lived in the dialectical movement between the two. I do not share Tuan's tendency to discuss these terms as if they were merely subjective experiences, but his distinctions do provide an introductory guide. If we start with space, it is experientially characterised by the fact that it creates the possibility of action. "Having space" can be understood both literally and figuratively. If I have space to put more books on my bookshelves, then I have the opportunity to acquire more books. If I have space in my home, I can also invite guests. If I have space in my daily life, I can also cultivate a new interest in abstract art. Space in the literal sense can sometimes be turned into space in a more metaphorical sense; in my examples of books, guests and art, although they move from the concrete to the more abstract, it will hopefully become clear how intertwined they are. Space on bookshelves signals not only a potential store of books but also an opportunity to expand horizons. According to Tuan, space – the open horizon – is a common symbol of freedom in Western society. Having physical space in your home or workplace also signals a certain freedom of action. The same is true of the larger physical space: "Space lies open; it suggests the future and invites action."[3]

2 Tuan 1977, p. 3.
3 Tuan 1977, p. 54.

Think of the meaning of the prairie or the expression "go west" in a classic Western film. Space does not belong, so to speak, to the familiar, but opens to the unfamiliar; it does not establish a "here", at least not directly, but rather points to a "there". Space signals departure rather than arrival. Although spaces are geographical in nature, here they do not merely denote geographical location; there is a significance to spaces and this significance is found precisely in freedom or in possibility.

Place is characterised, not unexpectedly, by its opposite. Place denotes the familiar, what we are used to, what we call "here", and thus represents an arrival or a stay rather than a departure. A prolonged acquaintance with a space often transforms it into a place that possesses a certain stability. We get to know the objects and spatial formations that give a concrete place its specific character, its appearance, sound, smell, even its flavour and tactile sensation. "Place is an organised world of meaning. It is essentially a static concept," according to Tuan.[4]

We get to know and get used to all the infra-ordinary details that make up the specific shape of a place, details that we can only become aware of through a deliberate recall, like Perec's and Søndergaard's attempts to exhaust or at least eavesdrop on Place Saint-Sulpice. The place assumes a meaning *for us* precisely by getting used to it, "inhabiting" it, letting it assume a specific definition and meaning. According to Tuan, the mother is the child's primary place, as she is "a focus of value, of nurture and support", but gradually the child conquers a more specific and geographical experience of place.[5] For the adult, place has acquired meaning as more and more memories are stored there, in furniture, walls, streets and squares. By conquering a space in the sense of making it less alien by getting to know it, we make it a place. This is not only a matter of sensory experience, but also a cognitive act, although the emphasis is on the former. As Tuan writes: "When space feels completely familiar to us, it has become place. Kinesthetic and perceptual experience as well as the ability

4 Tuan 1977, p. 179.
5 Tuan 1977, p. 29.

to form concepts are required for the change if the space is large."[6] Finally, the fact that a place is familiar to us also implies a certain permanence: I expect that the next time I come here, I will know my way around this place. This is also why it is possible, like Tuan, to associate place with security: the relative permanence of place both generates and is generated by trust.

Space opens up; place delimits.

A central task of architecture and urban planning throughout history has been to define – delimit – space, thus making it a place. Structuring space by means of walls, floors and roofs or walls, buildings, streets and squares creates the conditions for a place that is different from other places and where, by virtue of this difference, it is possible to find one's bearings and become familiar. The recurring debate about whether architecture or urban planning should open up our buildings and cities to the outside world or, on the contrary, define themselves in relation to it, illustrates the fact that the relationship between place and space is a tense but essentially dynamic one, rather than a mutually exclusive one. In a place without space we would feel trapped; in a space without place we would feel lost. However, place is not a static concept as Tuan argues, as long as it is dynamically connected to space. Instead, the experience of place is intimately linked to the ability to orient oneself there, as in the infamous test that prospective London taxi drivers are said to have to undergo: finding their way through the 15,000-plus streets and associated buildings and establishments that make up central London.[7]

Discussing the relationship between place and space, philosopher Karsten Harries begins by pointing out how the modern Western emphasis on space has emerged as a reaction against what he calls the "tyranny of place".[8] The traditional peasant was rooted to a specific place as destiny, a place that, in other words, the peasant never chose, but by chance and the weight of generations came

---

6 Tuan 1977, p. 73.
7 See Malpas 2018, p. 139.
8 Harries 1997, p. 168.

to be in. Modernity, not least through its technological changes, has increasingly meant that the ties to this specific place and its destiny have gradually become looser. The car has allowed the relationship between the home and the workplace to become more independent, thus reinforcing the separation between them in industrial society. Television, radio, telephone and the Internet have also liberated people from place, or rather allowed places to become increasingly interchangeable in post-industrial society. When our closest relationships can be maintained via social media without any particular regard to geographical location, we are seemingly no longer as dependent on our immediate surroundings and their specific characteristics. It is no longer the neighbours, the buildings and the surroundings but the peers on the Internet that constitute our primary social environment. We are detached, at least in our imaginations, in this and many other ways from the dependency on a concrete, delimitable place. Being less and less rooted in a specific place corresponds to a significant increase in freedom.

Here Harries finds a key to understanding the anomie and *ennui* of modernity: not being rooted anywhere means that the interchangeability and levelling of places is spreading. In a sense, he writes, "most of us today live in mobile homes"; our home can, in principle, be exchanged for another home at any time.[9] Against the "tyranny of place", therefore, comes the "terror of space".[10] In the infinite possibilities of space, we can't *live* anywhere in any real sense, which leaves us homeless, rootless, lost, adrift. For Harries, too, place and space are two inescapable aspects of human existence, with place having at least a temporal priority: the departure from something presupposes a previous position. A human being is both body and spirit, according to Harries, both earth and sky, but first comes the body and the earth. First comes place, then comes space.

Modernity's privileging of the infinite yet unbounded possibilities of space is not really an entirely new phenomenon.[11] The tendencies

9 Harries 1997, p. 172.
10 Harries 1997, p. 174.
11 See Casey 1998.

in Western philosophy since ancient Greece to seek and preferably find the essential and universal have favoured an understanding of spatiality as a homogeneous space, consisting of an infinite number of locations that are in principle interchangeable with each other and in whose empty space extended bodies can be precisely located. All the differences that exist in this space are understood as quantitative. At least from the Middle Ages onwards, space was imagined as infinite. Places, on the other hand, are qualitatively distinct, and therefore they are also heterogeneous, particular and, precisely because they are related to each other, they are also delimited from each other. They have character, nuances and history, sensory qualities that are only experienced by those who are actually physically there. They are "somewhere", unlike space, which is "everywhere".

From the perspective of place, existence is a *world*, in Greek a *cosmos*, or in other words a symbolic order between different regions or parts, rather than, as in the Latin, *universe*, a single whole.[12] A world is participated in, a universe is related to. Ultimately, infinite space, that is, the universe beyond the earth, becomes the epitome of possibility without limits, the space that more than anything illustrates the human dream of freedom as unlimited potential.[13] As a habitat, however, space leaves much to be desired; apart from the technical difficulties of colonising space, planet Earth remains, in an existential sense, home in most fictional depictions of space. "In space, no one can hear you scream", as the tagline of Ridley Scott's film *Alien* reads. Outer space is the epitome of a space that is "placeless", to use a term from phenomenologist Edward Relph – a term to which I will return in a later chapter – that is, a mute, impersonal space in which there is no resonance between the space and the occupant.[14]

Although I want to insist on the existential significance of space and not only of place – both living and moving are part of the basic

---

12 Casey 1998, p. 78.

13 See Harries 2001.

14 Relph 1976, cf. Asplund 2022, pp. 246–249. Here Asplund in some ways anticipates Rosa 2020.

human existential condition – we can now realise how the notion of abstract geometric space has developed out of the experience of space.[15] Having space means the possibility of imagining another way of existing than the concrete particularity of existence in a certain place. I don't have to live in Gothenburg, I could also live in Oslo; I don't have to work at the university, I could also work at a publishing house. By varying these ideas in the imagination, it is possible to imagine many possibilities. I could live in Paris, Rome, Oxford, Hong Kong, Tokyo, Stellenbosch and work in the wine business or for a football team or perhaps in an embassy. The variations of these opportunities are potentially endless. I could have residences all over the world and work with many different things regardless of where I am. If the emphasis is only on the possibilities, and these are allowed to obscure the concrete conditions, then we move, not only personally but also spatially, towards an ever greater degree of abstraction, until we reach the notion of a homogeneous space essentially without qualitative differences.

If the abstract notion of a single homogeneous space is allowed to dominate the life-world experience of heterogeneous spaces (the workplace, the home, the coffee shop, the concert hall, the traffic junction, and other mutually different spaces), then the significance of place for human existence will be overshadowed. Its concrete "somewhere" becomes irrelevant. Such a development is, however, not a fate, but in a sense a historical coincidence; the notion of the abstract, geometric space as the actual space is not an objective fact, but a historically evolved, socially constituted representation of a reified space.[16] By representing the relationship between human beings and space as a useful neutrality or a mute relationship between things, the representation hides how space is actually produced, what social relations it encompasses, and thereby also its history and meaning. But, as Henri Lefebvre notes, space as

15 See Guzzoni 1977.
16 See Lefebvre 1991; Landzelius 2009, pp. 39–75.

such is not homogeneous, it is only the abstract notion of it that has homogeneity as its goal and perspective.[17]

In fact, I cannot live anywhere in the world and do anything I want to do, for both economic and climatic reasons, but also because my life history, my social relationships, and my acquired desires and skills are not easily interchangeable. The experiences I have had in the places I have been have constituted the basis of the experiences I have now and thereby made me who I am, which in turn creates both constraints and opportunities for who I could be in the future. The interdependence between person and place constitutes a relationship that goes much deeper than the purely psychological or experiential and touches the very coordinates of our concrete existence. While a change in the specific relationship between person and space is always possible – I could both move and change jobs – flexibility is not unlimited. Even the traditional peasant Harris refers to always owned some kind of space, if you look closely enough, within the context of the place that has become the peasant's destiny, even if it is different from the space of modern mobility and flexibility.[18] To believe oneself to be completely independent of place is not only to forsake all continuity in one's own existence, but also to deny the horizon against which experience becomes possible at all.

Nor has the importance of place for human existence disappeared in modern times, but it tends to be obscured by abstract and generalising notions of space. Spatial practices – the way we move through space, inhabit it, experience it and appropriate it – can reveal an existential significance that is at odds with the representations of space that claim to tell us what it is. Despite claims about the levelling, place-dissolving practices of modern society, we still depend on places for our social relations, our relation to ourselves and to our memories.

17 Lefebvre 1991, p. 287.
18 See Asplund 2022, pp. 111–122.

Commonplaces

Places are social. Living together in a world means sharing things and places with each other, things and places that both unite and divide. The physical, material form of public, shared space is necessary for us to share it with each other. In her classic 1958 book on human action, *The Human Condition*, Hannah Arendt uses the image of a table to illustrate how this works. The table, she argues, is a symbol of living together because it is *between* those who sit around it:

> To live together in the world means essentially that a world of things is between those who have it in common, as a table is located between those who sit around it; the world, like every in-between, relates and separates men at the same time.[19]

The important word here is "between"; it is by virtue of its intermediate position that the table can both unite those who sit around it and separate them. Other things in the room function in the same way.

Arendt does not elaborate on the specific arrangement of this table, but presents it abstractly as a kind of necessary structure. Without it, if human beings lived only in nature without any given habitat, our existence would lack order; we would be homeless. Only through concrete objecthood, objectivity and permanence of things can the conditions for a common, in the sense of shared, world arise, a world in which we humans are able to act.[20] These conditions of human life, to which Arendt returns throughout *The Human Condition*, are indeed produced by humans, but nevertheless become the conditions of human existence. The common life that Arendt imagines is thus not only socially constituted through various relations between "I", "you" and "them", but rests on both a material and symbolic basis that in some sense precedes or at least is an indispensable part of social interactions. The external world necessary for communal life,

---

19 Arendt 1998, p. 52.
20 Arendt 1998, p. 7.

without which there would be no sense of reality, consists not only of other people but also of things, places, and their meaning.

In this shared world, we live together in the same way as we sit around a table, united but separated. You and I decide to meet at eight o'clock at one of central Gothenburg's meeting places and squares, Kungsportsplatsen. Our future meeting is constituted by virtue of a reference to a time but also a place that has a permanence that transcends our own subjective existence, a place that in a sense we are both equally as much and as little a part of. The site Kungsportsplatsen has a permanence and stability that neither you nor I can claim to possess. Either of us could forget our meeting, fall ill or, in the worst case, die. But still the very material condition of the meeting would exist. If, however, Kungsportsplatsen and other places were as ephemeral as a human life, it would be much more difficult to share a common world; if at eight o'clock Kungsportsplatsen no longer exists, we would not know where to meet. This may indeed be the case: the news is filled with daily reports of attacks and wars in which significant sites, historic neighbourhoods and entire cities are destroyed in Gaza or Ukraine. This is not only a material loss, but also a symbolic death: the conditions for this very common life are destroyed.

Common life is thus rooted in and dependent on the material world. One of the crucial distinctions for Arendt in *The Human Condition* is between the private life and the common life, where the former constitutes the sphere of necessity and the latter the sphere of freedom. Private life, too, has a material basis, but since the private constitutes that which is by definition detached from the public, it is not home and hearth that become the prime examples of the material conditions of the common world. Life together is underpinned by public buildings, streets and squares that have a permanence and stability that transcends a single generation. Although communal life presupposes private life, where place and property are intimately linked, unlike in private life, communal buildings, streets and squares are not the property of any particular person. Such property is shared in a sense that does not coincide with legal ownership. It would be strange to suggest that we should meet at *my* Kungsportsplatsen at eight o'clock, even if it were the case that I have a deeper relationship with Kungsportsplatsen than you do by virtue of Gothenburg being home to me.

If life were to be lived only in private, we would not be seen and listened to by others, nor would we see and hear others ourselves. Nor would we relate to each other through the mediation of things where, just as around the table, we are both united and separated. A merely private life would be a life of alienation. The purely private life is, if not an inhuman life, at least a life deprived of something essential to humanity, namely

> to be deprived of the reality that comes from being seen and heard by others, to be deprived of an "objective" relationship with them that comes from being related to and separated from them through the intermediary of a common world of things, to be deprived of the possibility of achieving something more permanent than life itself.[21]

In other words, it is through the detour of things that we can have an "objective" relationship with each other and share a world. Regardless of whether Kungsportsplatsen has a greater – or substantially different – personal significance for me than for you, Kungsportsplatsen is an "objective" place in relation to both of us. In Arendt's words, Kungsportsplatsen constitutes such a "common world of things" in which human encounters, actions and positions can be "seen and heard". It creates a space where human actions can, as one says, "take place". Place enables an encounter between the self and the other.

How does Kungsportsplatsen become a real place and not just a place that helps us know where to meet? From Arendt's further reasoning, we understand that for her it is about sharing a world, that is, a symbolic order that includes the material as an indispensable component, and not just about sharing a specific place. That the world we share, according to Arendt, consists not only of locations, but above all of places, is explained by her argument about the "masses" in an earlier book, *The Origins of Totalitarianism* (1951).

21 Arendt 1998, p. 58.

What characterises the masses above all, according to Arendt, are their "essential homelessness".[22] The masses are collectives for those individuals who have lost the ties to their fellow human beings – in the sense of citizens – that are mediated by a common, shared world of things. Other terms Arendt uses to describe the homelessness of the masses in *The Origins of Totalitarianism* as well as in *The Human Condition* are "alienation" and "rootlessness".[23] The term "homelessness" points particularly well to the experience of no longer having a given place in the world. For Arendt, "homelessness" is not necessarily about having nowhere to live, but above all about having no given abode, no "immediate earthly surroundings".[24] Arendt describes this condition as being a stranger to the world, even an intra-worldly alienation from the world. The alienation in question here is thus not primarily having nowhere to live, but the loss of a world shared with others. What Arendt's homeless person lacks is not just a place that can be called home in any genuine sense, but existentially significant places in general. Both the private and the common world disappear in the experience of a general placelessness.

The loss of common places, of a shared world of things, also means the loss of a *sensus communis*, a "common sense". In German, the term Arendt uses is "*Gemeinsinn*", so the *sensus communis* is not primarily about what we sometimes call "common sense", that is, something like the "general opinion".[25] Instead, "common sense" is about "shared reason" based on sensory impressions and exchanges between people who share a world from several different perspectives. The "atrophy of the space of appearance" therefore has as a consequence "the withering of common sense, the organ with which we orientate ourselves in it".[26] This is why homelessness, alienation from the earth and the world, also results in the loss of a

22 Arendt 1976, p. 352.
23 See Arendt 1998, pp. 248–257; Arendt 1976, pp. 196 f., 232, *et passim*.
24 Arendt 1998, p. 251.
25 Arendt 2021, pp. 395–402; Arendt 1998, pp. 280–284.
26 Arendt 2021, p. 296; Arendt 1998, p. 209. I here follow the German edition.

genuine *sensus communis* in the sense of a reason shared with others. Instead, we are driven back to our own self-reflection, which has no content other than our own streams of consciousness, and where our own thought itself becomes the foundation of the reliability of our own existence: "common sense [...] now became an inner faculty without any world relationship".[27] This "common sense" consists only in everyone reasoning alike, not in a world common to all. Therefore, the reliability of the outside world, the world we share with each other and with things, becomes increasingly precarious, since it does not possess the kind of apodictic certainty with which Cartesian reason is associated, nor is it shared by all of us in any existential sense. There is no longer, metaphorically speaking, a table around which to gather when each of us is left to their own devices. When our reason is reduced to logical propositions and our experience of the world is relegated to the merely subjective, then the very notion of a "shared reason" becomes obsolete. As human beings we become individuals, we become lonely, confined to ourselves; a private being in the etymological sense of the word, after the Latin *privus* meaning "separate".

The lack of space gives birth to loneliness, which in turn gives rise to conformist masses. The individual in the masses is the lonely being, a human being defined by lack of meaningful social relations. When the "shared reason" of the *sensus communis* is transformed into a "common sense", deprived of all interaction between people and things and transformed into what "everyone" thinks, then we also lose our ability to orientate ourselves in a common world. Loneliness, being rootless and therefore strictly private, is the breeding ground for totalitarian movements, according to Arendt. "To be uprooted means to have no place in the world, recognized and guaranteed by others", because it is such a place that makes opinions meaningful and actions effective.[28] Faced with such loneliness, faced with the chaos that must characterise the world made up of nothing but individual atoms, the totalitarian

27 Arendt 1998, p. 283.
28 Arendt 1976, p. 475.

temptation in its various forms, with its promise of order and roots, becomes almost impossible to resist and even to contradict.[29] But this promise of order consists again not in a divided world with its mixture of the disordered and the ordered, but in uniformity. There is no longer a shared world with common places, where a shared reason could appear valid.

Arendt's discussion of the loss of a genuine *sensus communis* as a consequence of the loss of common places raises questions about the importance of place for human subjectivity. Arendt does not, of course, mean that the individual in the masses literally leaves the earth or the world. The individual does not suffer a loss of their location but rather of their place, when there is neither a place that is home for them nor, above all, a place shared with others. Another term, rootless, is used by Arendt, as well as by Simone Weil in the same post-war period, but for neither of them is the need for roots about an ethnic destiny associated with a particular geographical location. It is, rather, about the meaningful belonging to a particular place. To have roots is to have a natural residence in a collective community and a hometown, Weil argues, but it does not bind us to a single place: "Every human being needs to have multiple roots."[30] The experience that Arendt and others want to put their finger on is not about the disappearance overnight of the streets, squares and buildings that have surrounded people, even though many of these were wiped out when European cities were bombed during the Second World War. What is lost is the experience of these places as communal spaces, as commonplaces. Without such places, human existence becomes discontinuous and constantly flexible, which is also the breeding ground for totalitarian movements.[31]

29 Arendt 1976, p. 352.
30 Weil 1952, p. 41.
31 Arendt 1976, pp. 306, 316 f., 352, 478. Cf. Sennett 2000.

The Memory of Place

Arendt's reasoning about common places is insightful, but rather abstract. Apart from the parable of the table, we learn very little about what set of things characterises a place. Is it possible to be more concrete?

A common way of thinking about how a place becomes a place is to see it as the site of a historical event. In many cases, of course, this is entirely correct, but the relationship between place and person becomes even clearer if we think of memory (here distinct from history) as the third pole of the relationship. The memory we are talking about here is not just about a specific event, but relates to a broader sensory experience of a place. What moves me about a place is not just what once took place there, if anything in particular happened there at all. The sense of place, its smell, sound, sight, tactile qualities and perhaps even flavour, as well as our embodied existence all play a part in the memory of a place, just as Perec and Søndergaard have shown with regard to Place Saint-Sulpice. The experience of space is multi-sensory and so is the memory of place. It evokes memories in us that are not only images of the past but also reminiscences of how we once related to a place, including to those who have populated the place with us, even those who are now dead.[32] When the memory of our childhood home, or our old school, returns to us on a later visit, it is a first-hand bodily experience, not a secondary experience, like looking at a photograph from the past. Essential to this memory is that we actually visit the place. It is well known since antiquity that physically visiting a place, as opposed to merely hearing about it or reading about it, creates a more vivid memory. If the place, for example our old school, were to be demolished, then the conditions for such a memory would also disappear.

32 On being with the dead as part of collective memory, see Ruin 2018, especially pp. 34 f. Ruin's philosophical discussion of the dead and especially the places of the dead in relation to the memory of the place is profound in relation to the memory of the place in a way that I cannot do justice to here but which I hope to return to.

But what kind of memory is this? After Henri Bergson, a distinction has sometimes been made between recollective memory, the act of remembering something, and habit memory, in which repetition is the central feature.[33] This distinction will not be explored further here. Suffice it to say that, although place memory touches on both, its association with habit memory is central, especially as we may generally tend to ignore this dimension of memory in favour of thinking of memory as recollection rather than as bodily habit. But in fact, it is not only with the mind that we remember, but with the whole body. "The body stabilises memories through habituation and reinforces them through the power of affect", writes Aleida Assmann.[34] Memory is thus not only cognitive, it is also affective and habitual: it concerns our emotions and habits. When we talk about memory of place, it is primarily the affective and habitual that come to the fore. Emotions and habits also link us to other people, to what is usually referred to as collective memory.

Ever since Maurice Halbwachs' theory of collective memory from the first half of the 20th century, it has been clear in much of memory research that our memories are not only individual, but both rooted in and generated by the social group we belong to as well as by our physical surroundings. Halbwachs begins his posthumously published book *The Collective Memory* (1950) by referring to the experience of returning to a city we have previously visited and allowing the memories we have forgotten to be awakened by what we see.[35] To vary his observation, we can also recall the experience of the demolition of a building, again our old school in our home town, and with it the memories, of a more or less personal kind, not only associated with it but also generated by it.[36] As Halbwachs points out, such place-bound memories are reinforced if we also share them

33 See Ricœur 2004, pp. 23–26 *et passim*.

34 Aleida Assmann 2018, p. 20.

35 Halbwachs 1980, p. 22. As a background to this posthumously published work by Halbwachs from 1950, see also Halbwachs 1992, originally published in 1925. Further aspects of collective memory can be found in Jan Assmann 2007 and also Aleida Assmann 2018, pp. 130–145.

36 See Sigurdson 2024a, pp. 23–30.

with each other through conversation or when they are conveyed to us through texts; for Halbwachs, Charles Dickens' novels came to orientate him on his first visit to London. All the exchanges we had with our classmates in our school also shaped and preserved our memories of the places in the school where they took place. For Halbwachs, collective memory is not only social but also rooted in concrete places.

Halbwachs' primary interest is the role of the social group in collective memory. He points out how our ability to remember is so linked to a social community that the loss of this community means a loss of memory in the sense that we no longer have access to certain memories: "To forget a period of one's life is to lose contact with those who then surrounded us."[37] Our belonging to such a memory collective is not just formal, but a concrete participation that is affectively charged. Memories are cultivated here, as they are passed back and forth between the members of the memory collective in conversations and reflections. Such acts of reminiscence can be effective precisely because they are based on a life shared between the members. If I do not share my existence with the group in question for a long period of time, if I do not share a community and thus no visions, hopes and affects, then I will not be able to recall the memories we used to share to the same extent. I no longer remember what I remember from the perspective of the group. It also means that I no longer have volitional control over these memories in the same way as before; I cannot, as they say, "recall" what happened from the shared horizon of the collective. These collective memories, Halbwachs argues, are not the sum of individual memories: "The collective memory [...] encompasses the individual memories while remaining distinct from them."[38] Of course, I don't remember anyone else's memory in collective memory; it keeps our individual memories distinct from each other. But collective memory works in a more "external" way than individual memory; it provides a framework for those individual memories that are more "internal".

37 Halbwachs 1980, p. 30.
38 Halbwachs 1980, p. 36.

Moreover, for Halbwachs, memory and history are two different things. Only when collective memory fades or is destroyed does history begin. Memory presupposes the continuity of the memory collective, and since collective memory is associated with particular groups, there are several such groups. But when they cease, so does memory. History, on the other hand, is interested in discontinuities, but it is only one, not several. In a way, history presupposes that the memory collective has ceased.

It is not my task to question such a sharp distinction between memory and history here. Nor is the point to determine the validity of a memory, but rather its function in relation to place. Let us therefore turn to the importance of place for collective memory. Halbwachs is well aware of the difference between geometric space and existentially significant place. Things, he argues, give us a sense of order and permanence. Our physical environment bears the imprint of ourselves and of others. Things are part of human society. It is through things that the collective also gets a sense of its own stability:

> The place a group occupies is not like a blackboard, where one may write and erase figures at will [...] But place and group have each received the imprint of the other. Therefore every phase of the group can be translated into spatial terms, and its residence is nothing but the juncture of all these terms.[39]

Above all, it is the cities, their neighbourhoods and the buildings of the neighbourhoods that represent such permanence. Halbwachs, who spent much of his life studying or teaching in Paris, is of course well aware of Baron Hausmann's regulation project between 1853 and 1870, but Paris along with Rome are nevertheless mentioned as examples of cities that have seemingly crossed the centuries with preserved continuity, despite wars, revolutions and crises.

It is worse for the memory collective when a particular significant street is remodelled or a particular building or home is demolished. The memory in question seems to be more associated with places of

39 Halbwachs 1980, p. 130.

neighbourhood dimensions. The groups we are part of are associated with very concrete places, like the cobbler with his workshop, and not just the city as a whole. Of course, the permanence associated with the city is not absolute: collectives evolve and the material configuration of cities changes, "[b]ut even if the stones are movable, relationships established between the stones and men are not so easily altered".[40] A place will adapt itself, so to speak, to the habits of its long-term inhabitants, but their habits – the tendency to think certain thoughts and move in a certain way – will in turn be organised by the material configuration of a place. The physical object of a place becomes an image of the local tradition. Its boundaries also become the boundaries of tradition and sometimes, as in pre-modern cities, of their jurisdiction. If memory is to be understood as something permanent, as something shared, and as something that has the capacity to recall the past to the present, or to sense how what has been lingers in the present, then it is because place and memory are so intimately linked that they can be seen as aspects of each other.

Halbwachs here gives a hint of how place relates to memory. Places give memory a permanence it does not possess in its individual form. As Paul Ricœur writes: "These places 'remain' as inscriptions, monuments, potentially as documents, whereas memories transmitted along the oral path fly away as do the words themselves."[41] The memory of place is not primarily about particular memorials that in some way memorialise something that happened at the place in question, perhaps in the form of a monument or statue, perhaps in the form of a particular name that recalls some person or event, as when Napoleon named two hundred streets in Paris to glorify his name and achievements.[42] Nor is it about individual memories of a place. Rather, it is the memory of the place itself as an embodiment of collective memory.

40 Halbwachs 1980, p. 133.

41 Ricœur 2004, p. 41. Ricœur also provides a detailed critical discussion of Halbwachs' idea of collective memory on pp. 120–124.

42 See Connerton 2009, p. 12.

Is it possible, then, to speak of such a memory of place without ascribing to it a consciousness of its own? Halbwachs was far from such thoughts: the collective memory was embedded in what he calls a "social frame", not a collective unconscious. And it is possible to go further than Halbwachs in bringing together human activity and spatial conditioning in terms of memory. For Arendt, the obvious fact that people construct the streets, squares, and buildings of cities does not mean that this fact deprives these structures of their role as a precondition for the human life that takes place there. They establish a system of places and spaces that exist independently of us and before us, even if previous generations have constructed them and we must inevitably be in the process of reconstructing them. Time adds a quality to places that eludes our presumptuous claims to be their immediate cause. In places, the past lingers in the present as something that has been.

We can easily see that different urban designs give rise to different ways of being: narrow streets without sidewalks imply a different way of moving than wide streets with sidewalks; squares provide space for gathering for collective action while car lanes separate block from block; building facades can be more inviting or more confining.[43] The way we design our cities through architecture, based on what we think we need or what we consider to be an aesthetically pleasing cityscape, also gives rise to habits that adapt to or are based on this configuration of space. Moving around a familiar city is based on a certain habitus. In this sense, the material configuration of space is a form of memory: habit embodied in stone. According to Richard Sennett in *Flesh and Stone*, our late modern cities are characterised by an individualistic notion of human communication that borrows its model from William Harvey's discovery of the constantly pulsating bloodstream.[44] In a constantly circulating flow, the public places – such as squares – where people stop, meet, and interact, tend to disappear, because they do not correspond to the individualistic conception of the human being. Instead, busy streets

43 See Sigurdson 2024b, pp. 85–96.
44 Sennett 1994.

will become the type that will shape modern cities, and therefore how we move around in them. My point is not to dwell on this particular problem, but to let it become an illustration of how Sennett understands the relationship between bodies and cities: not only at the level of the imagination, but also in the habitus embodied in our patterns of movement. The shape of the city will affect our existence. Our habitat is shaped by our habitus and vice versa.

That memory is dependent on place is not a modern realisation. On the contrary, it has been actively cultivated in the rhetorical art of memory since antiquity, for example by Cicero:

> people desiring to train this faculty must select localities and form mental images of the facts they wish to remember and store those memories in the localities, with the result that the arrangement will preserve the order of the facts, and the images of the facts will designate the facts themselves, and we shall employ the localities and images respectively as a wax writing tablet and the letters written on it.[45]

This kind of memorial art, which flourished for many centuries until the breakthrough of the printing press, i.e. for more than 1,500 years, is known as the "loci method" – after the Latin *locus*, i.e. place – or a memory palace.[46]

Here the memory palace, which may be imaginary or a real building in one's mind's eye, is furnished in the mind of the memory artist with images associated with what is to be remembered. The various parts of the palace – its entrance, halls, rooms, staircases, inner rooms, columns, and so on – can all be used as places in which images are placed, and then when the memorialiser's mind returns to a particular place in the palace, it will again encounter the image in question. If there is a lot to be remembered, many different places are needed. The locations can also be affectively charged, as affectivity favours memory.

---

45 Cicero 1967, 86: 354.

46 See Yates 2014; Spence 1985. The loci method is of course not the only form of memory art in antiquity; on the scriptural culture of early Judaism as a cultural memory technique, see Assmann 2007, pp. 212–228.

The memory palace as an art of remembering is based on a few basic premises: firstly, that the various places in the palace form an ordered and stable system, and secondly, that the act of remembering takes place on a human scale.[47] For many of the thinkers who formulated the rhetorical art of memory, the sense of sight was also in the foreground. It is therefore not surprising that, in the Middle Ages, the art of memory came to be embodied in physical and not just mental buildings: for example, medieval Gothic cathedrals can be understood as a kind of memory palace made of stone. This ancient art of memory really works; history is full of examples of memory artists who, using this technique, mastered for us unimaginable amounts of detailed knowledge.

In a way, this is reminiscent of ideas we have already encountered in the work of Hannah Arendt and Maurice Halbwachs: if common places, that is, the *loci communes* of rhetoric, are a prerequisite for both shared reason and collective memory, then it is not very far-fetched to think that places can also be used to practice memory.

If habitus and habitat are linked, we can imagine, following the suggestion by rhetoric, that concrete, physical places also function as wax tablets on which our memories are inscribed. Or, to use the other image of memory that has dominated history, places function, according to St Augustine, as "treasuries" or "storehouses" for memories that they reveal during our visits, or "hiding places" where they are "stored away available for retrieval".[48] Thus, on a collective level, memory and forgetting would be intimately linked to the built environment itself and our ability to orientate ourselves in it. The disappearance of our common places, either as a result of the increased homelessness or placelessness that Arendt in particular diagnoses in modern society, or simply because certain places are demolished or destroyed in war, is also a kind of memory loss. Common memories depend on common places. It is of course possible to try to recreate such shared places in memory – it is precisely this ability on which the art of memory is based – but however important such memory

47 See Connerton 2009, p. 5.

48 Augustine 2006, X, 8, 12; see Aleida Assmann 2018, pp. 149–178.

work may be, it will never recreate the concrete places themselves as such. We can no more visit Paris before Hausmann's regulation than we can visit the mental memory palace of the memory artist Matteo Ricci, and therefore we will never be able to experience the Paris that is now called the "romantic Paris" as a living experience, only as a bygone memory.[49] Unlike the memory palace of rhetoric, where sight is at the centre, such a living experience of place is multi-sensory. In principle, it involves all the senses.

There is a fundamental difference between what are usually called places of memory and what I am talking about here as the memory of place. In the modern urban space, we are surrounded by memorials that refer us to a distant or recent past: cenotaphs, statues, mausoleums, monumental works of art of various kinds. However, as Pierre Nora has pointed out, the inclination toward memorials in the modern city is rather an effect of the disappearance of our common memories: "There are *lieux de mémoire*, sites of memory, because there are no longer *milieux de mémoire*, real environments of memory."[50] Using a French pun, Nora contrasts *lieux* with *milieux*. A memorial can never replace place as an environment for memory. "Memorials conceal the past as much as they cause us to remember it", writes Paul Connerton in the spirit of Nora, and we can add that they can never bring forth the experience generated by the past.[51] As an act of remembrance, they relegate what is to be remembered to the past as much as they make it present to the present.

The place I am talking about here in the context of place memory is more like Nora's "environments of memory": a memory sedimented in stone, wood or other material that also often bears traces of its production process. Cities, neighbourhoods and buildings can be understood as a kind of mnemonic device at different scales. If the place of remembrance demands our attention, the memory of the place makes itself felt through our inattention, when we take it for granted. The memory of the place is an environment for memory:

49 See Marrinan 2009.
50 Nora 1989, p. 7.
51 Connerton 2009, p. 29.

"Memory takes root in the concrete, in spaces, gestures, images, and objects".[52] In the text quoted here, Nora makes a strong thesis about how history is contrasted with and replaces memory. But the question is whether the memory of place does not make history and memory more familiar with each other, at least from the perspective of experience. And the even more crucial question is how inattention can be crucial to memory.

At Kungsportsplatsen stands the statue of the king Charles IX of Sweden. This king founded a city that was a precursor to today's Gothenburg, a city that was also called Gothenburg and existed between 1603 and 1611 before it was burned down by the Danes. Today's Gothenburg was founded ten years later, in 1621, by Gustavus Adolphus, the son and successor of Charles IX. The statue of Charles IX was inaugurated in 1904 but moved slightly in 1936, although its location is still on the street Östra Hamngatan at Kungsportsplatsen. At the beginning of the 20th century, it was felt that Charles IX deserved a statue precisely for his contributions to the predecessor of the city of Gothenburg, and a memorial was therefore established at Kungsportsplatsen. Nowadays, few people probably recognise the king on the statue and know anything about why he is standing here. Instead, the statue is known as the "Copper Mare", which refers both to the material this statue is made of and the horse Charles IX is sitting on.

The statue serves as a central meeting point.

It is, to use Connerton's words about what I refer to as the difference between a memorial and the memory of a place, not so much an object that we look at to remember a past event, but rather "something which is inconspicuously familiar to us" and which can therefore evoke memories in those of us who are in the habit of passing or meeting at Kungsportsplatsen.[53] Rarely, if ever, do I reflect on Charles IX and his early Gothenburg; instead, the statue becomes part of the gestalt of the place, the background that makes me realise, even in my inattention – or precisely because of it – that

52 Nora 1989, p. 9.
53 Connerton 2009, p. 34.

I am in Kungsportsplatsen and not somewhere else. The statue of Charles IX has gone from being part of the extraordinary design of Kungsportsplatsen to becoming part of its infra-ordinary design, precisely in Perec's sense: that which is so ordinary that it lies beneath the ordinary. If one were to write a book about Kungsportsplatsen congenial to the character of the place, it would not be a history of Charles IX and his Gothenburg, but rather an "attempt to exhaust and listen in on a place in Gothenburg" in the wake of Perec and Søndergaard. In other words, the kind of memories generated by significant places are firstly habitual memories, closely associated with affective memories, and only secondly cognitive memories in which we recall a particular historical event. It is the kind of inattention that comes from my habitus, an inattention that comes from repeatedly being in the place rather than thinking about it, that makes the Kungsportsplatsen a place for me, a place that is partly mine.

Cities, neighbourhoods, buildings and monuments can act as mnemonic devices for us. But this very idea must not give us the illusion that it is our own human activity as such that generates memory in this case. In fact, it is the place that awakens the memory in us, the memory that is sedimented in the place. Another difference with the art of memory is therefore that the concrete places can surprise and amaze, because, unlike the places of rhetorical memory, they are not arbitrarily chosen. The concrete place is thus in a certain sense given. Place Saint-Sulpice in Paris or Kungsportsplatsen in Gothenburg are places by virtue of the fact that they have become the places they are for us through all those who have built, used, visited, lived in and with these places for a long time. We appropriate them by using, visiting or living in them ourselves. In the habit of visiting these places, we acquire the disposition to experience them as places, as familiar places – as in some sense *our* places. This is where a continuity between the place and our actions and affects arises. What makes the Place Saint-Sulpice in Paris and Kungsportsplatsen in Gothenburg places that are meaningful to us in some everyday sense is simply that we revisit them.

To understand more deeply how places work, we need to turn to the question of the relationship between place as something we discover and place as something we create. Is there any real contradiction here?

Place is not only about how we share commonplaces with each other, how place is part of our collective memory and how we become part of this memory through place. Place is also about how we use and experience a place in and through its stones and atmospheres, i.e. its material form as well as its affective tonality. If place is not just a blank canvas onto which we can project our preferred conceptions of it, since it always already enables certain attitudes, uses and experiences rather than others through its concrete form, then it also becomes relevant to argue that we are part of place even before it becomes what we make it.

Place and person do not relate to each other as object and subject. The dynamics between them therefore need further investigation. How do we allow place to become place by returning to it, even repeating it? What does it mean that place in some sense objectively enables certain uses and experiences? What is the significance of the classical notion of a *genius loci* – the spirit of place – for our understanding of place as found rather than created? How can we understand the everyday atmospheres as given, despite the bias towards manufactured, constructed places that characterises our time? To inhabit a place, to relate to a place as place, presupposes an experience of place in which place and person cannot be separated in any simple way. To claim that place is first and foremost found or given must therefore not exclude that it is also, in some sense, created or constructed. This chapter will explore this dialectical realisation.

Repetitions

A man goes to Berlin and has a memorable experience of the city. The room he stays in during his trip has a fine view of a moonlit city at night, the visit to the theatre where he sees a farce is enchanting and in the audience sits a young woman on whom he rests his eyes. The experience of Berlin was so satisfying that he decides to make exactly the same journey once more. But nothing will be the same as last time. He doesn't get his usual box at the theatre, the audience is disruptive and the young woman is nowhere to be seen. His room is different, his morning coffee doesn't taste good, and neither does the food at the restaurant where he usually eats. The next day, when he returns to the theatre, he is faced with the same disappointment. His return to Berlin was not what he had hoped for. He is unable to repeat it. "The only thing repeated was the impossibility of repetition."[1]

This little story is taken from Søren Kierkegaard's book *Repetition*, one of the most important works for the reflection on repetition as a form of knowledge since its publication in 1843. Kierkegaard demonstrates through this concrete example that an identical repetition is not possible, the only repetition that is possible is a non-identical repetition. For the second Berlin journey to have been more successful for the book's protagonist Constantin Constantius, that is, for repetition to have succeeded as repetition, he would have had to undertake this Berlin journey as a *new* journey. If a place becomes a place for us through repetition, which is essentially what I argued in the last chapter, then this happens through a constantly varied return visit or reuse. For a place to be constituted as a place, even a memorable place like Berlin for Constantin Constantius, then this presupposes a dialectic between the familiar and the new, where on the one hand the place must *have been* in order to be repeated, but on the other hand *becomes new* precisely by being repeated.[2] This is

1 Kierkegaard 1964, p. 75. For a modern rehearsal of Berlin, see Jordheim 2023.

2 Kierkegaard 1964, p. 52.

how our memories of places become memories that affect us in the future, through repeated revisits.

Kierkegaard is well aware of this. His alter ego in *Repetition*, Constantin Constantius – literally the doubly unchanging – serves as a contrast to one of the insights the book wants to convey, namely that permanence can only be achieved through change as non-identical repetition.[3] Place is maintained as such not only by looking back at what it once was, through retention, but at the same time also by imagining what it can now become, through anticipation. In order to truly preserve Berlin in our memories, we must allow Berlin to meet us anew, in a partially different form, on our return visit. Or, to vary the place in question, we must adopt the same perspective as the person in Selma Lagerlöf's novel *Jerusalem* who did not go to Jerusalem after all but remained in Nås in Dalarna; the very intention of leaving home transformed this familiar place into something new in a non-identical repetition.

In a sense, therefore, it is true that place is something we ourselves create through repetition. Places are created through continuous use. Kungsportsplatsen is clearly a place constructed by people and for people. Once upon a time, someone drew a street here, houses were built, a statue was erected. Not overnight, but over years, decades, centuries. It was only in 1852 that it was named after the city gate that stood here until 1839. But the very fact that it was located at a city gate means that it has been something of a crossroads since the 17th century. Back then, there was no fixed bridge over the moat, instead there was a drawbridge to ensure communication to the south. Like other similar sites, it adapted to the changing city and technological advances: no tolls are collected anymore, tram and car traffic flows more or less unhindered. Its character gradually changed, from control centre to traffic hub.

Kungsportsplatsen became the place it is today through an ongoing process of collective construction, a non-identical repetition, which of course continues even today. In the present, it has become a place for me, and not just a position on a map, through my constant return

3 See also Pickstock 2013.

visits, on my way past or waiting for someone I am going to meet there. In this way I get used to Kungsportsplatsen; I take in all its ordinary and its infra-ordinary characteristics through my senses, and learn to recognise them so that they increasingly become subliminally perceived signals that I find myself at Kungsportsplatsen. People have been travelling here since the 17th century, passing through on their way to somewhere else, or simply lingering in anticipation of someone else. A place like this is not just made of stones. It invites people to use it in different ways, combining materiality with memories and function. The stones themselves become a memorial to its constant reuse.

Of course, my contribution to this collective construction process that makes Kungsportsplatsen a place is minimal. I use it. I pass it on my way Gustaf Adolfs torg (a square named after Gustavus Adolphus), meet people here, or hop on tram number 5 on my way home. I remember it in the way I described in the previous chapter; in this way it becomes part of my personal psychogeography, the orientating map of Gothenburg's inner city that I carry with me and which is more experiential and bodily-habituational than mental. Experiential because it establishes a relationship with the concrete places I visit and which themselves generate experience in me; I do not need to recall streets and places in my mind through an act of memory for the map to work. I become part of Kungsportsplatsen through use and experience, and the only sense in which I myself establish the place is by repeating the use and experience of the same place by others. Above all, I have nothing whatsoever to do with its material construction; I have not created it other than through the – surely non-trivial – co-creation that takes place through my (re)use of the place. I approach Kungsportsplatsen as a *fait accompli*, admittedly changeable, as history has shown, but still fairly stable in the long term. Not much in life is more permanent than the stones of the city.[4]

For me personally, it is more correct to say that I discover Kungsportsplatsen than that I create it. This is also the case in practice

4 Marcus 2023, pp. 33 f.

for everyone else. Of course, there are urban planners, architects, construction workers and others who were once involved in building the stones that make up the site, but this construction is a collective act. It may be correct to say that it is we as people who construct Kungsportsplatsen but that most of us are not involved at all, a few very little involved and even fewer more involved. But even those who once constructed the bridge and laid down the tram tracks or decided that a bus stop should be located here, did not encounter Kungsportsplatsen solely as a work of their own agency. For them, too, their own use and experience of the emerging site was affected. This is not primarily because they were not the only ones involved in the construction process, but because of the inescapably external nature of the emerging and completed work. Every use is a re-use, even the first. Therefore, every visit is a return visit, including the first.

We encounter a place as something other than an immediate outflow of our labour. We approach it as a material, functional and atmospheric reality outside ourselves, even though we were directly involved in its construction. The fact that Kungsportsplatsen is a work of the human hand does not contradict the fact that when it was built it took on characteristics independent of us. Those who once built Kungsportsplatsen moved stones and created a shape. But the paving stones are a processed raw material and these are arranged into a form that draws inspiration from previous similar sites, that takes some account of the physical circumstances that characterise the position where the site is to be built. And finally, the site assumes an independent character to which further construction must likewise relate. Perhaps it is more accurate to say that a place like Kungsportsplatsen is constructed through *processing* rather than *creation*, given the theological associations of the latter word? Although intentional processing presupposes an agency on the part of the person doing the processing, it also presupposes a kind of autonomy and inertia on the part of the material being processed. Processing takes place in relation to the possibilities offered by what is to be processed, both building materials and the topography of the site in question; it is not a creation out of nothing. It is impossible to completely master what is to be worked on; instead, it is a matter of discovering what possibilities are contained in what is to be worked on. Here too, human agency consists primarily in the ability to respond.[5]

This is how we usually relate to places, beyond the concrete construction of them, as something we discover rather than something we construct. It is probably a mistake here to understand creating and finding as mutually exclusive alternatives; discovering and later getting to know a place also means, in a weak sense, becoming part of its continued existence in the form of a non-identical repetition. When I visit Kungsportsplatsen or travel to Berlin, I also contribute, in some minimal sense, to their continued existence. The point, however, is that the place is primary and my visit secondary. Places, as I said, have a permanence that surpasses my own. Therefore, they will also shape me more than I shape them. As urban morphologist Lars Marcus puts it: "We could say that it is not about the people and their places but about the places and their people."[6] This means that it is the places that invite me, that provide the conditions for my use and experience of them, and in this way process me, or who I am at that moment. This happens in a way that is largely outside my own control and consciousness. Their form, which I discover when I visit them, establishes a habitus in me whose origin is outside myself.

Enablers

To better clarify the impact that places have on us, we can turn to the psychologist James Gibson and his concept of *affordance*, which frames the relationship of animals to their habitat or ecological niche. According to Gibson, the *affordance* of the environment is "what it *offers* the animal, what it *proposes* or *furnishes*, either for

---

5 By extension, we can associate this with post-humanist reasoning on the relationship between agency and response, for example Bennet 2010 or Haraway 2016.

6 Marcus 2023, p. 140.

good or for ill [...] It implies the complementarity of the animal and the environment".[7]

Different environments have different affordances in relation to different animals. The same is also true in relation to humans.

A simple but clear example Gibson uses is a surface that is "horizontal, flat, extended, rigid, and knee-high relative to a perceiver" and therefore appears to be user-friendly for sitting.[8] Not only is it physically possible for a person to sit on, it also looks welcoming, it invites people to sit down. Its affordance, the way it suggests we should sit down, can be perceived visually. Human beings are sitting animals, and since sitting is part of our repertoire of possible actions, this surface appears sit-friendly *to us as humans*, but not necessarily to dogs, who sit in a different way, or to birds, who prefer to sit on a branch because their knee joints do not allow them to bend their legs in the same way as we do. Affordances are, as Gibson writes, both subjective and objective; being both mental and physical characteristics, they point both ways and show why these dichotomies are inadequate for understanding how affordances work.[9] "Sitability" is an objective property of certain surfaces, in the sense that it is possible to physically sense it as independent of us, but at the same time a relative property because it presupposes a certain physical constitution in us (knee joints in the centre of the legs) to be meaningful. Our senses – not just sight – guide us to perceive the affordances that a given environment offers based on the beings we are. Doors invite us to enter or exit, floors to stand up and move around, windows to look out or in, stairs to descend or ascend, and so on.[10] Therefore, Gibson can argue that we are created by our environment in a more fundamental way than we create the environment ourselves. Air, water, earth and fire in their various forms are, so to speak, the elements of our environment which,

7 Gibson 1986, p. 127.
8 Gibson 1986, p. 128.
9 Gibson 1986, p. 129.
10 Cf. Pallasmaa 2017, p. 133.

although we can process them in different ways by actually relating to them, are nevertheless inescapable.

Affordances are "enablers".[11] The affordances that characterise a particular habitat or place make it possible to use it in different ways based on the experience of its physical form. A human being, a dog and a bird can use a "horizontal, flat, extended, rigid" surface in different ways depending on their different physical constitution. But even between people, the same place can allow different uses of the place. Human-constructed habitats do not determine our actions, but can make some uses of a particular place more possible and others less possible to varying degrees. For example, building a football pitch on a sloping meadow is less functional. Marcus uses playing football on a sloping field as an example of how there are always inherent tendencies in those enablers that characterise a particular concrete place: "the game goes on but over time it gives the game a certain direction".[12] Sure, you can use a place or thing in a way that goes against what it primarily enables – I can stand on the chair instead of sitting on it – but certain inherent tendencies of what the place or thing enables privilege certain uses more than others, give it a certain direction and way of living together with others. The sloping football pitch, over time, gives an advantage to the team playing downhill.

Places can thus be understood as a particular kind of enabler (or conglomerate of enablers). What they make possible naturally differs from place to place. In its current form, Kungsportsplatsen is a hub for traffic, especially local traffic in the form of buses and trams, but also to some extent car traffic and also, of course, pedestrian traffic. At Kungsportsplatsen, I can choose to take a tram southwards, in the direction of Götaplatsen, a central square in Gothenburg, but also northwards, towards other such places, Brunnsparken and Gustaf Adolfs torg. If I choose to walk, I can, for example, head towards Järntorget by following along the channel westwards or towards the square Drottningtorget and the Central Station by

11 Marcus 2023, p. 23.
12 Marcus 2023, p. 139.

following it northwards. Kungsportsplatsen has a relatively high degree of centrality, as it enables several different connections with other places. At the same time, it is also built to allow for many different uses. Not only do I have to pass it on my way to somewhere else, but I can also stop there at a café or at the statue, waiting for someone I know to pass by. There are a number of shops, a market hall nearby and some open spaces. Although it is a busy place, there are sections with characteristics that make you want to linger there. Other busy intersections such as Korsvägen or Brunnsparken have fewer such opportunities, depending on their specific shape. They have gradually become more efficient from a public transport point of view, their accessibility and reach have increased, but by being specified for a particular function they have also lost some of the traditional hub's character of enabling a wider range of uses. They have become increasingly neutral, increasingly instrumentalist, increasingly monofunctional in character.

And so far I have only focused on traffic nodes in the centre of Gothenburg, but in fact there are countless different places in a city, each with its own unique possibilities. A train station, a church, a library, a park, a neighbourhood with apartment buildings or one with single-family homes, spaces left over after planning, industrial premises, sparsely trafficked streets, all constitute different kinds of spaces in the city that carry their own unique possibilities.[13] As Marcus writes, the shape of space "has a great impact on our relationship with other people, limiting and giving direction to our freedom of action".[14] The way spaces – that is, the building, the site, the neighbourhood, the city – are constructed affects not only how we relate to the space in question, but also how we relate to each other.

If we compare, in our minds, the open city park that houses a number of cafés with the busy metro station, we realise that these different places reward different postures, walking rhythms, ways

13 See Sigurdson 2024c, pp. 127–152 and 2024b respectively. See Marcus 2023, p. 33.

14 Marcus 2023, p. 33.

of meeting the gaze of others and generally relating to other people. The street favours a higher level of mobility than the walk between the houses in the neighbourhood. Add to this the idea of the longevity of these structures by virtue of their built form, and we can see how they have a long-term impact on how we relate to them. Just as it is possible to play football uphill, it is of course also possible to go against the flow in most spaces, figuratively and literally, but in the long run a certain habitat will favour one habitus over another. "Slowly but surely, the building gets its way."[15]

Terroir*: A Digression*

Another example of how place objectively enables different uses, approaches and consequences can be found in the French wine term *terroir*: "the total natural environment of any viticultural site" according to the standard work *The Oxford Companion to Wine*.[16] The main components of *terroir* are not only soil, but also topography and climate (macro, meso and micro), the combination of which gives each site its unique character, which also leads to a more or less consistent wine over the years. "Thus every small plot, and in generic terms every larger area, and ultimately region, may have distinctive wine-style characteristics which cannot be precisely replicated elsewhere."[17] According to wine-drinking philosopher Pierre-Yves Quvigier, *terroir* "refers to the land, to the geology, to the sunshine, to the rainfall, to the nature of the slope".[18] Grapes grown on clay soil do not produce the same wine as grapes grown on volcanic soil.

To become wine, grapes must be grown in a specific place (or several) at a specific time.[19] But of course, wine quality is not

15 Marcus 2023, p. 106.
16 Robinson and Harding 2015, p. 737.
17 Robinson and Harding 2015, p. 737.
18 Quivigier 2023, p. 39.
19 See Quivigier's phenomenological determination of wine in Quivigier 2023, p. 62.

only dependent on *terroir*, but also on variety of grape, producer, vinification process, vintage and pure chance. As a classification concept, it is not universal or always the most interesting. Nonetheless, Burgundy wines from different sites have distinctly different quality and style, despite being grown and vinified in the same way.

Terroir should therefore play a significant role in the quality of the wine.

Scientific studies have found that a common denominator for top wines is that the soil in which the vines grow has a balanced fertility and a moderate water supply, so it is not extreme in either direction. Climate also plays an important role, as does topography, which means that soil is one of several factors that determine the quality and style of the wine. At the same time, *The Oxford Companion to Wine* points out that the increasing number of wines made from grapes grown on a single vineyard from countries other than France, particularly the New World, indicates that *terroir*, and especially soil, is recognised by many winemakers as a crucial factor in viticulture.[20]

Nicola Perullo emphasises that a wine's *terroir* should never lead us to believe that the place it refers to is something stable. On the contrary, *terroir* is always something that is constantly being re-created by the work put into growing the wine as well as by the natural conditions: how even the soil and climate in which the vines grow is a dynamic reality that we ourselves are part of. "A *terroir* is a question that provokes experience rather than a predetermined answer; it is not the cause, it is a process."[21] Of course, wine is dependent on time (vintage) as well as place (*terroir*), but neither of these characteristics is an objective label by which wine can be classified. The character of a wine does not *come from* its *terroir*. It does not *express* a *terroir*. It reflects a *terroir* in a way that cannot be captured by terms like cause and effect. In other words, *terroir*, although it exists in a concrete way, is not a static reality. "The wine of a *terroir* is a process of growth – where birthplaces, things, and

20 Robinson and Harding 2015, p. 739.
21 Perullo 2021, p. 166. Originally italicised.

people continuously combine – that corresponds with the world."[22] In other words, according to Perullo, *terroir* is not a predictive factor but an enabler – an *affordance* – for the surprise that is the specific, unique wine. In this way, Perullo wants to avoid both claiming that *terroir* does not exist and perceiving *terroir* in objectivist, static terms. *Terroir*, for Perullo, like place in general, is something that *happens* rather than something that *is*. All places, including wineries that produce wines that are not to anyone's taste but also places that have nothing to do with winemaking, have their *terroir*.

But making wine is as much about repetition as it is about surprise: winemaking is an experiential skill that involves extensive insights into soils, topography and climate. The fact that certain vineyards regularly produce memorable wines gives the winemaker reasons to believe that this can also happen in the future, reasons based precisely on such insights, as well as on his craftsmanship. The fact that it is possible to act on these insights with a certain confidence of repeatability should mean that *terroir*, even as a living reality, has a constancy that means that it not only *becomes* but also *is* or *is* precisely because it is constantly *becoming*. Perullo's emphasis on the experience *of* wine as something impossible to objectify risks leading to too sharp a contrast with a knowledge of winemaker, vintage and *terroir*, as if these must be mutually exclusive. As G. W. F. Hegel notes, repetition and spontaneity are not necessarily opposites.[23] The same, of course, applies to places in general, not just *terroir*: they *are* precisely by being constantly repeated and therefore acquire in and through this repetition a relative continuity and stability over time. *Terroir*, like places, *is* precisely in that it is constantly *becoming*, and what *terroir* specifically *becomes* is an enabler for specific, unique wines.

---

22 Perullo 2021, p. 172. Originally italicised.
23 Hegel 1979, pp. 187–191.

The Atmosphere of Place

The possibilities we find in a place are not limited to a number of different options for action. At Kungsportsplatsen, I do not need to look out over the place with my eyes to orientate myself to any great extent, because I know it so well through habit. Instead, all the countless infra-ordinary expressions of the space will meet me as subliminally perceived synaesthetic impressions that guide my specific posture. My body interacts with the possibilities of space.

This interaction between space and body is intertwined with an affective charge of the space. Its atmosphere affects how I orientate myself in it: "To sense our own presence is at the same time to sense the space in which we are present."[24] This does not mean that the atmosphere of a place like Kungsportsplatsen has to be particularly charged or overwhelming; on the contrary, the feeling of a traffic hub can often be characterised by a low-intensity atmosphere. Nevertheless, the pragmatic enablers of place are always inevitably intertwined with atmospheric enablers. The ways in which places evoke certain actions rather than others are not purely physical, like the sloping football pitch, but in, with and through those physical attributes, they are also affective. Atmosphere is not something extra to the physical form of the place: the atmosphere of the place reveals its affective character to us, and is intimately connected to its material form – the expectation of the train station, the dull muteness of the city corridor, the cosiness of the home, the vertical sanctity of the cathedral, the hospitable intimacy of the restaurant, the exclusivity of the shop, the uncanny mood of the cemetery after dark, and so on. In doing so, atmosphere also shows us how to behave or conduct ourselves there. Without this affective charge of some kind, our places would not be places in the true sense, but only spaces. To the same extent that we are surrounded by space, we are also surrounded by its atmosphere and the affordances it provides. Therefore, we also encounter the atmosphere of the place not primarily as something we

---

24 Böhme 2013, p. 110.

create ourselves, but first and foremost as something we discover in the place, as something that prompts us and claims us.[25]

Genius Loci

Tim Ingold has described two different, complementary ways of relating to our spatial existence as inhalation and exhalation: we breathe in the atmosphere of the place and breathe out its structure.[26] In the experience of atmospheres, there is no strict distinction between subject and object. The fact that we also participate in the production of places does not contradict the fact that we are first immersed in the atmosphere of the place, like the swimmer in the water, and can therefore take a few strokes forward. If the place we are in is really a place and not just a locality, then as living, sentient, corporeal beings we always already have some kind of affective relationship with its atmosphere.

This may sound far-fetched, especially in our time, where place is often reduced to position and where agency is constantly emphasised over experience. Who is emotionally affected by visiting Kungsportsplatsen?

Any place, however monotonous or dull, can be described by adjectives that capture something of *how* it appears to us. Its atmosphere becomes an *affordance* for certain moods, certain ways of affectively relating to a space, as well as – and in close harmony with, even inseparably intertwined with – the material design of the place. As Gibson writes: "The perceiving of an affordance is not a process of perceiving a value-free physical object to which meaning is somehow added in a way that no one has been able to agree upon; it is a process of perceiving a value-rich ecological object."[27]

A classic expression of the impact of place on humans beyond the purely physical or material is the notion of a *genius loci*. The "spirit of place" stands for an experience of being seized by an external

25 Cf. Griffero 2020, pp. 99–114.
26 Ingold 2015, p. 87.
27 Gibson 1986, p. 140.

power associated with a concrete place. In the *Aeneid*, Virgil (70 BCE–19 BCE) describes how Aeneas worships the "spirit of place" (or the "area's local spirit", as my translation has it) – as well as other divine beings associated with place – when he first arrives in Latium (which eventually becomes Rome) to pay his respects to the place:

> He garlands his temples with leafing sprigs. Then, addressing the area's local spirit, invoking Earth as the first among gods, then the Nymphs, and the rivers not yet known, Night and the night-rising stars, he now offers his prayers, Jupiter, Ida's lord, comes next, then the Phrygian Mother, both or his two parents then: one in heaven, one in Erebus' darkness.[28]

Some places, or perhaps even most places, were considered to be animated and it therefore became the task of human beings to come to terms with the spirit of the place in order to be able to inhabit the place in question. Aeneas could hardly have got away with less, since his purpose was to settle in Latium.

Today, the term is widely used in architecture and phenomenology, and in popular accounts of the impression made by a particular place. Aldo Rossi (re)introduced *genius loci* in his 1966 *The Architecture of the City* and its importance for understanding architecture's relationship to culture and place; Christian Norberg-Schulz uses *genius loci* precisely as a starting point for a phenomenological investigation of how the architect can build meaningful places that help people live; Vernon Lee (pseudonym of Violet Paget) invokes the spirit of place in her travelogues, where she is more interested in the inner impression places convey than their material configuration.[29] My own interest here is primarily phenomenological, assuming that the spirit of place as a phenomenon can be illuminating for our experience of places as places even today, and that a certain historical background can then also shed light on our experience of

28 Virgil 2007, p. 162 (7, 136–140).
29 Rossi 1984, pp. 103 ff; Norberg-Schulz 1980; Lee 2021; Blennow 2024.

place and how it has changed profoundly over the centuries. Hence this historical sketch.

The Latin term for the spirit of place, *genius loci*, was established in the Augustan period (63 BCE–14 CE), but the thing itself is older than that.[30] Some places held people spellbound: a nymph sanctuary, where beauty plays an important role, old and mighty trees perceived as sublime, groves of trees that overwhelm the visitor, places perceived as terrible. Some places had a numinous atmosphere that left its mark on the visitor. The concept of *numen*, divine power, has both an impersonal and a personal side: both an impersonal spirit and a personal spirit, one might say. The spirit of the place could be experienced both as an impersonal atmosphere of a concrete place and as an epiphany, a concrete beholding of the nymph or demon that inhabited or was identified with the place. In the Greek language there is no technical term for a *daimon topou* or anything similar. Instead, they speak of divine or sacred places and what these places have in common is that something demonic inhabits them – in the ancient Greek sense of *daimon* as a kind of divine power or spirit, not in the Christian as an evil power. In the Augustan period, the concept of *numen* increasingly comes to denote something personal, concrete and place-bound, although Roman religiosity was more distanced from these. There is therefore a rich mythical background to the concept.

In particular, at that time, *numen* referred to natural places that had not yet been touched by human hands, where one could experience the spirit of the place: caves, hollows, springs, trees, rivers, mountains and so on.[31] These were approached with reverence by assimilating or submitting to them rather than trying to master them. Even in the case of human-made places, such as the Roman Forum, which was considered to be the *umbilicus* of the earth, its navel, the main concern was to shape them in such a way as to express the character of the already given, the basic form of the place. In no way were

30 For a comprehensive account of the cultural history of the *genius loci* concept, see Kozljanič 2004a; Kozljanič 2004b.

31 See my discussion of the cave of the nymphs in the first chapter above.

these places seen to express the spirit of human beings; the task of the edifice was to correspond to the numinous power of the place. Even when the Olympian gods placed the archaic, place-bound gods in a wider context, the concept of *genius loci* was preserved by subordinating them. Zeus and Hera, Poseidon and Demeter, Apollo and Artemis, Ares and Aphrodite, Hephaestus and Athena, Hermes and Hestia all had a more universal character, but were still, often, perceived as belonging to a particular landscape or place, albeit not as inseparably as the spirit of that place. (In Virgil, Zeus is associated, as we have seen, with the Cretan mountain Ida.) The various *genii locorum* took their place within a broader numinous landscape, to which human beings were also to relate reverently.

However, the beginning of Christianity meant a transformation in the understanding of the spirit of place. The Christian church's concept of creation meant that God was thought of as more radically separate from God's creation than in Olympian mythology, and therefore in principle not connected to concrete places at all. Its eschatology also meant that the spirits of the various places were met with great scepticism; either they were perceived as evil spirits or they were transformed by being represented as local saints, good people associated with a particular place. Nor is there a Christian technical term for the spirits of place, although they survive in popular religion. According to Robert Kozljanič, there is a cautious and critical approach to this whole complex of ideas, even though grey areas are developing between the ancient mythical and the Christian, for example in the form of haunted places and places of pilgrimage. Conceptually, experientially and architecturally, then, Christianity represents a more radical transformation of the notion of the spirit of place than Olympic mythology, in two main ways. Firstly, in the way it moralises by dividing the spirits into evil and good – demons and local saints – where previously they were more ambivalent. Secondly, by making the place an allegory for something that transcends the place itself. Kozljanič writes: "The earthly-numinous place is superimposed by the otherworldly-allegorical place, the sacred tree by the Christian cult building".[32] The concrete locality of the place, he argues, became in principle secondary.

Yet Christianity does not in all respects break with previous conceptions and experiences of the spirit of place. Although it represents a radical transformation of ancient thought, among other things by emphasising how God transcends any concrete place, there are also, at the same time and in the same Christianity, more centripetal forces, in the form of the localisation of God's *logos* to a concrete person and place – above all in Jesus of Nazareth – and the related experience of the sacramentality of the created world, that is, its God-given disposition to manifest and mediate divine presence.[33] God was not to be sought alongside creation, but in, with and under it. Thus, there was also a theological incentive to regard certain concrete places as meaningful and potentially epiphanic. Hence, well into the early modern period and perhaps beyond, there were notions of the significance of particular places visited by apostles or saints, of places where visions took place or the Eucharist was celebrated, of places where monasteries and churches were built as monuments to significant memories, places that carry their own atmospheres and memories. Kozljanič's thesis about the allegorical superstructure of the earthly-numinous places only makes sense if it is not interpreted as a radical discontinuity to the Olympian religion that comes before. Rather, because of its emphasis on the earthly and created nature of man, lived Christianity clarified the existential significance of concrete places, while also transforming the meaning of these places in relation to ancient religion.

In any case, by constituting a historical caesura, Christianity could pave the way for a more radical break with the ancient idea of a spirit of place. The modern notion of the spirit of place is no longer

32 Kozljanič 2004a, p. 379.

33 Here, more than Kozljanič, I want to emphasise Christianity's dual relationship to place: not only centrifugal but also, more than ancient religion, centripetal; not only topophobic, but also topophilic. For divergent perspectives on this issue, see Casey 1997 and Sigurdson 2016, respectively. Kozljanič's presentation here does not appear to be incorrect, rather one-sided. See Gassenhuber and Kozljanič 2024, pp. 22, 74 f.

of a spirit inseparably associated with the place, but of a spirit that expresses a person's subjective impression of the place. Kozljanič again: "Here the ancient concept of *genius loci* becomes an aesthetic metaphor. The daimonic place becomes a metaphorical place."[34] It is thus the human being itself who gives the place its spirit; the spirit it finds in the place is none other than its own spirit. When Petrarch, as one of the forerunners of the Renaissance, climbs Mont Ventoux on 26 April 1336, he is not looking for an ancient *daimon* or a local saint.[35] What interests him is his own experience of the place. The ascent of the mountain becomes a metaphor for something that happens within Petrarch himself. Thus the spirit of the place is transformed into the spirit of the subject, a spirit that is in principle independent of the place. It is no longer the place in itself that forms the subject, but the subject freed from all places that allows the place to become a concrete metaphor for its inner mood.

Thus, the function of place comes to be an external reason for seeing oneself, but in doing so it also becomes more or less disenchanted. Although places and landscapes may continue to be experienced aesthetically as sublime, beautiful or picturesque, the aesthetic experience is not understood as meaning that these places or landscapes have any atmospheric or numinous dimension in their own right, whether of the "daimonic" or demonic kind. Place is merely a resource, it becomes what we make of it; lived space must therefore also give way to geometric space. Human being's relationship to place has changed radically compared with antiquity. If, in archaic antiquity, a human being fitted itself into the place, and in Olympian antiquity reshaped the place without deforming it, Christianity's superstructure of the place eventually results in the modern planning relationship to the place.[36] It is now the human being as such who disposes of the place through planning, who makes it a place and not just a position. In this division between subject and object, there is no doubt where the agency lies, namely

34 Kozljanič 2004b, p. 15.
35 See Kozljanič 2004b, p. 17.
36 See Kozljanič 2004b, p. 258, for an overview.

in the human subject, but the ancient and medieval correspondence between place and person is thus in principle broken.

In the modern world, too, there are accounts of overwhelming experiences associated with places, experiences that border on the numinous and inescapable and are thus not just a capacity drawn from humanity's own range of possibilities, especially as regards the atmospheric.[37] Romanticism, in particular, has been a culturally significant force in reclaiming a language for such experiences. But even here it is sometimes uncertain whether the spirit belongs primarily to the place or to the subject, whether our sense of a place means that we are really gripped by its atmosphere or whether we merely savour at a safe distance the impressions it produces in us.[38] For contemporary humans, however, the socially mediated imaginative horizon consists of a fundamental distinction between subject and object and a geometrical space without qualitative distinctions between affective registers (position rather than place). Thus, the articulation of experiences of atmosphere, place and spirit of place tends to be reduced to the subject's feelings about a place, even when the experience of grasping or being overwhelmed by the place actually transcends this division. The spirit of the place then becomes a metaphor for my private mood, no longer a force associated with the concrete place that I have to relate to. The place becomes something I – or rather we – create, not something that is found.

The Atmosphere of the Everyday Place

After all, it is perhaps not a given that experiences of the spirit of place, or at least of the atmosphere associated with a particular place, can always be relegated to private experience, even in our own time. Experiences of entry, where I am overwhelmed by the atmosphere of the room, as well as experiences of discrepancy,

37 See e.g. Eliade 1959; Otto 2014.
38 See Schmitz 2019b, pp. 641–649.

where the atmosphere of the room differs from my own mood in a way that sheds light on both, are common.

Atmospheres are thus not something that is generated by the experiencing subject without residue. They are anchored in the spaces that surrounds us. The encounter with an atmosphere implies an encounter with something other than myself, something that has the potential power to change my mood.[39] It may be a change in the weather, the entry into a reverently quiet room from a noisy street or a crowded football stadium, or simply the everyday recognition of revisiting Place Saint-Sulpice or Kungsportsplatsen. What they have in common, as I said, is not the intensity of these experiences of the atmosphere of the place, but the priority of the atmospheres in relation to the people who experience them.

But how given is the priority of such a place? Based on the idea that places come into being through repetition – reproduction – Paul Connerton has questioned any talk of the "'the power of the place' or of the 'genius loci'" as a fetishisation of place that hides the work that actually goes into producing them.[40] In fact, it is an atmospherically gifted architect, a few marketers and a local construction company that have produced the atmosphere of a Paris square rather than a spirit of place. It is therefore through labour that we bring about a permanence of places that allows them to become "the locus of institutionalised power and culturally specific perception".[41] Marxist-inspired philosophy, in particular, has insightfully pointed out that the real process is concealed in favour of a mystifying talk of a nature-given power when the very production of place is made invisible by talking about its spirit.[42] One of the activities of ideology criticism is to reveal how the supposedly natural is in fact a cultural construction. Indeed, those in our own time who advocate that there are still genuine experiences of a spirit of place usually point out that

39 Cf. Kozljanič 2004b, p. 276; Bollnow 2009, pp. 37–43; Schmitz 2019a, pp. 361–367.
40 Connerton 2009, p. 51.
41 Connerton 2009, p. 51.
42 Harvey 1990; Lefebvre 1991.

this does not mean that these always are a given, a purely passive reception, but that the relationship to place is a form of labour. It requires an effort to come to terms, including architecturally, with the place you want to inhabit.[43] Nevertheless, it is easy to imagine how talk of a spirit of place can serve a chauvinistic ideology that wants to refer, for example, the characteristics of the population to a basis in the concrete place.

The atmosphere that rises up from a particular place and appears to be natural may in fact be socially generated.

If the spirit of place, here conceived primarily as a kind of atmosphere rather than a concrete and personified *daimon*, is nothing more than a projection of the meaning-making subject (in individual or collective form), such a reference of the affects to the social subject seems to presuppose both a strict separation between human subject and place object and an equally strict reduction of space to a homogeneous geometrical space. However, the fundamental experience of embodied existence and spatiality in humans cannot be likened to a third-person perspective that relates to the place as a spectator; the lived space, which for the experiencing human is the primary thing, is experienced through the sensory impressions that we make when we are physically there, in place, in the first person.[44] The lived space is not the same as the observed space, because the latter presupposes a distance between subject and space that is secondary, in purely experiential terms. Our primary experience of space is therefore not how we regard the room, but an immediate, pre-reflexive grasp of it. It is an experience that takes place, at least in part, on a preconscious level. We realise, for example, that we are at home, without really having to think about it.

The lived space, which is the infra-ordinary space I explore in this second part of this book, is also not a homogeneous space. Instead, it is made up of qualitatively different meanings and values that differ from place to place and between which there are thresholds to cross. It does not correspond to the abstract, geometric, homogeneous space

43 Cf. Norberg-Schulz 1980, pp. 10 f. *and passim*.
44 See here Bollnow 2011, pp. 17–25.

and is not even, in our experience, primary. As Henri Lefebvre in particular made clear, the notion of geometric, homogeneous space as actual space is by no means self-evident as space actually is, but a historically developed and socially constituted representation of space.[45] Space as a homogenised resource, as a pure position, is a "culturally specific perception" that can therefore conceal "the locus of institutionalised power" as Connerton warns. An ideology-critical perspective on place therefore points in two directions: on the one hand, against the talk of the spirit of place as a way of concealing the conditions of the production of place, but on the other hand, against the notion of the affective neutrality of space as a mystification of the both naturally and historically evolved atmosphere that inevitably characterises the lived experience of a place. If so, it seems a philosophically legitimate task to investigate whether a *genius loci* in a particular place is found or created.[46] Moreover, it does not have to be one or the other, unless we assume a radical separation between subject and place as well as a homogeneous space. Such a perspective does not exclude that some alleged *genii locorum* may be subjective projections, but it presupposes that the lived, bodily experience of space is a genuine expression of human existence and that there are therefore several possible relations between subject and place.

In line with his idea of three different ideal types of atmospheres – prototypical, derived and inauthentic – Tonino Griffero has suggested that these ideal types can be applied also to our interaction with places. To the prototypical atmospheres, which are "objective, external and non-intentional", corresponds a relationship with the place in which we completely surrender to its atmosphere; to the derived atmospheres, which are "objective, external and intentionally generated", corresponds a relationship with the place in which we intervene in and reshape its atmosphere, like the Christian reinterpretation of former holy places in the quote by Kozljanič above; to the inauthentic atmospheres that are "subjective

45 See, again, Lefebvre 1991.
46 See Griffero 2020, p. 143.

and projective" corresponds a relationship to the place in which it is entirely we ourselves who plan and shape its atmosphere.[47] Here it becomes clear that the position of the subject and the place in relation to each other can shift and is not static. In the first case, the direct experience of the spirit of a place is greatest, and in the third, where human agency is emphasised, there is both the possibility of a genuine unmasking of an inauthentic spirit of place and the risk of an illusory perception of a homogeneous place, reduced to a resource and completely emptied of specific atmosphere. To speak, as Norberg-Schultz does, of a "man-made *genius loci*" in contrast to a "natural *genius loci*" is, in Griffero's view, a contradiction in terms.[48]

In the city, where I mainly find myself, which is clearly a social, human construction, it is above all options two and three that are relevant. Especially with regard to option two, it becomes clear that the atmosphere of the city cannot be reduced to either the subject or the place, but rather presupposes an interaction between the two. The city cannot, in Griffero's words, be "thought of as a static contemplative 'object' but also as a constructive circulation between the given (felt) and the configured (enacted)".[49] That "place is never a fixed spatial entity but always a social process in transformation", as Connerton argues, therefore does not contradict Griffero's rejection that every atmosphere is "always and only" the "result of an intentional project".[50] The fact that human beings through repetition become part of a place and therefore does not merely remain passive-receptive does not mean that they radically

47 Griffero 2017, p. 28; cf. Griffero 2021, pp. 88–105; cf. Griffero 2020, p. 144.

48 See Norberg-Schultz 1980, pp. 58, 69, 77; Griffero 2020, pp. 140 f., 143, 145.

49 Griffero 2020, p. 144; cf. Thibaud 2015, pp. 58 f.

50 Connerton 2009, p. 51; Griffero 2020, p. 144. In Martin Heidegger's classic text "Building Dwelling Thinking" (Heidegger 1975, pp. 145–161; German original Heidegger 2000, pp. 147–164), it is never a question of the human act of "dwelling" being unilaterally about our active "setting up" but always also about our passive "allowing".

create that place. It is actually more accurate to talk about how they "inscribe" themselves, or perhaps even "weave" themselves and their history into the already existing fabric of a place.[51] If there is indeed a spirit of place, at least in the form of a specific atmosphere, it can only be produced within certain limits, because it is primarily found.[52] If the spirit of place is not first something we are given, it cannot become something we help to create. This has often happened in history: a specific sense of a *genius loci* gives rise to an incentive to build on the site in a way that reflects this sense, as in the case of the Roman Forum.

It is not surprising that the impression arises that we human beings are the sole and unique producers of atmosphere, especially in our time, with its emphasis on human agency. But the fact that the spirit of the place, wherever it exists, does not wholly compel us does not mean that it does not exist and operate in the form of atmosphere. To begin with, as I said, it is part of the contemporary socially mediated imaginative horizon that affective experiences belong to the subject and can only be attributed in a metaphorical sense – or as an illusory projection – to a place in some sense. Even when the concrete experiences contradict such a division, it is such a perspective that has interpretative primacy. Furthermore, the spirit of the place, at least in its less assertive form, is conveyed by modest, diffuse and infra-ordinary sensory impressions belonging to the "holistic affective tonality" of the place.[53] These are so low-key that they risk remaining unnoticed by those who do not relinquish their agency in a kind of passive, receptive, almost pre-reflexive attention, for example by giving themselves over to the place in the way we initially saw Perec do with Place Saint-Sulpice. Perec "notes" things rather than "stares" at them. The lived experience of places does not consist of a number of sensory impressions combined with a sense of where I exist in relation to everything else that is then synthesised, but consists of a totality of impressions, where both subject and place

51 Kozljanič 2004b, pp. 405 f.
52 Griffero 2020, p. 145.
53 Griffero 2020, p. 144.

are involved in this tonality in a way that has not yet separated them. *How* we live the place does not distinguish between pragmatic and affective affordances or enablings in any simple way.

Even such a relatively simple and unglamorous place as Kungsportsplatsen offers not only a number of possibilities for action, but also – in and through and with these – moods as part of its environment. For the casual visitor, who may call Gothenburg home, such an unassuming place can also serve as an identification with the wider city and an existential point of orientation. Norberg-Schultz writes: "To belong to a place means to have an existential foothold, in a concrete everyday sense."[54]

Kungsportsplatsen can be several things at once: a hub for local travel in Gothenburg, a place to meet, but also perhaps, for citizens of Gothenburg, an everyday reminder in condensed form of an affective relationship with their own hometown.

54 Norberg-Schultz 1980, p. 23.

La Défense

La Défense is a district in the Île-de-France, three kilometres west of the centre of Paris.[1] It takes its name from a sculpture by Louis-Ernest Barrias erected in 1883, *La défense de Paris*, in memory of the soldiers who defended the city in the Franco-German war a decade earlier. The bronze sculpture consists of three patriotic figures executed in an Art Nouveau or "Jugend" style. It now contrasts sharply with the office landscape and commercial centre that is today's La Défense. The neighbourhood's history only began in 1958, when a decision was made to try to revive the old neighbourhood of factories, farms and substandard housing. Today, La Défense is not only the commercial centre of Paris, but also an educational centre with no less than four business schools and other universities. Here the visitor finds a cluster of skyscrapers, with as many as fourteen reaching more than 150 metres from the ground. The two tallest, Tour First and Tour Hekla, measure 225 (excluding an antenna) and 220 metres respectively, making them among France's tallest buildings. 180 000 workers a day visit the district, far more than its inhabitants. The skyscrapers surround a huge, wide esplanade, the Esplanade du Général de Gaulle, which seems fairly empty at most times of the day, except at lunchtime when there are many people milling around the 150 or so restaurants, most of which are inside the shopping centres. It is also home to Europe's largest indoor arena, with 40 000 seats, a giant exhibition hall that looks like a triangular tent stretched across a large glass façade and now houses the Centre national de l'industrie et des techniques, a vast

1 Much of the information about La Défense can be found on its official website "Paris La Défense, bien plus qu-un quartier d'affaires" (2025).

and futuristic multi-storey indoor gallery. There are around 2 600 hotel rooms in the neighbourhood.

La Défense is no longer a poor and neglected neighbourhood on the outskirts of Paris.

During my visits to La Défense, I have wandered back and forth along the esplanade, and made a few short excursions to neighbouring sites. The word esplanade probably conjures up thoughts of a walkway lined with vegetation, but although there are many trees flanking the Esplanade du Général de Gaulle, it is easy to overlook these, as the large surface area of the walkway itself dominates the north-western part of the esplanade. At the south-eastern end, however, is "Le parc", a "jardin lineaire", where neat rows of trees and other plants create a less harsh impression of the esplanade.

My attempt to find a café where I could sit and look out onto the esplanade and wait for lunch was more difficult than I thought. Eventually I found a Starbucks where I could look out at all the glory. Eventually I also saw an "Italian trattoria" – named just that, in English and Italian – in the middle of the wide street. There are plenty of cafés and eateries – in the shopping centres. Almost none face the esplanade. Watching the surroundings and all the passers-by from the viewpoint of the café becomes, unlike in central Paris, an activity fraught with difficulties. There are certainly green ambitions at La Défense, where an expanded park could make the district carbon-neutral, but despite the plantings, its environment is characterised by a distinctly urban atmosphere. The glass of the skyscrapers reflects other skyscrapers far above each treetop. The solitary stroller, with no destination other than a café from which to view the place, easily experiences themself as very small, wandering among the wide expanses of the esplanade. La Défense is the place to go if you have a purpose for your visit, not so much to look around, although there is plenty to look at. The artistic decoration is extensive and many buildings are iconic and award-winning.

La Défense is built to exude modernity, especially in the sense of striving forwards and upwards. Apart from being surrounded by skyscrapers, the esplanade is not made of traditional cobblestones, but of moulded gravel slabs, ground or unground. It is the right angles that dominate the area, not natural and irregular formations. Most striking, and probably a reason why many tourists still come to a district like this, is the Grande Arche de la Fraternité, a monument in the form of a cube-shaped triumphal arch 110 metres high, wide and deep, which houses a large number of offices.[2] Built on the initiative of President François Mitterand between 1982 and 1989, the "Great Arch of Fraternity" consists of a concrete frame covered in glass and granite (the Cararra marble cracked and was replaced). It was designed by Danish architect Johan Otto von Spreckelsen and constructed by Danish engineer Erik Reitzel. Intended to inspire humanitarian ideals, rather than to celebrate military victories, it was nevertheless inaugurated, somewhat ironically, by a military parade in July 1989, on the 200th anniversary of the French Revolution. Under a wall-less roof that looks like a sailcloth stretched horizontally in the centre of the arch itself is a kiosk in the form of a steel box advertising the panoramic view from the cube's roof. But on each of my visits to the neighbourhood, there is literally a note on the door, not because it is sold out, but because access to the rooftop view that existed in the original plan is permanently closed. It doesn't seem particularly easy to cross the threshold of any of La Défense's buildings apart from the shops and restaurants. So I have to content myself with walking up the imposing staircase, to the elevation where the Grande Arche sits, and standing under the arch. Its entire interior is made up of windows arranged symmetrically in groups of seven by seven, but it seems impossible to work out what might be going on behind them, even though I know that there are 37 floors of offices with a total floor area of 140 000 square metres. Instead, I turn to look out towards La Défense and Paris.

In addition to all these magnificent buildings, another important detail that adds to the monumentality of the Grande Arche is the four-kilometre straight line that runs from the Louvre through the Arc de Triomphe to the Grand Arche, the "axe historique du Paris". Admittedly, the Grand Arche is twisted 6.33 degrees compared to the

2 On the Grande Arche, from the perspective of its structural engineer, see Reitzel 1984, pp. 191–197.

historical axis, but standing in the centre of the arch, the view to the Arc de Triomphe is unobstructed and a breathtaking perspective of Paris unfolds before your eyes. Framed by skyscrapers and the grey tiles of the ground, it seems to the eye that the esplanade extends unbroken to the Arc de Triomphe and beyond. The Grande Arche is thus not a break with the monumentality of central Paris or France; on the contrary, it follows a continuous historical axis going back centuries. If La Défense is today the main business centre of France, designed in a resolutely contemporary modernist, international style, this does not mean that any significant historical caesura has taken place in relation to historic Paris, at least historic Paris after Baron Haussmann's regulation. Together with the Montparnasse Tower, a secondary axis is formed with the Eiffel Tower in between, where these two, both some of the tallest buildings of their time in Paris, are joined by the Grande Arche. The historical and geographical continuity with the forward-looking, upwardly mobile Paris can be considered quite extensive.

As a kind of counterpoint to the monumental – to say the least – character of La Défense, a church and a sculpture can be mentioned here. On a side branch of the esplanade is the almost brutalist church of Notre-Dame de Pentecôte (architect Franck Hammoutène), which opened after the turn of the millennium.[3] The church itself is beautifully designed, in an austere manner, with corrugated iron, raw concrete, wooden seating and an altar and a baptismal font in treated iron. Aesthetically, the church fits in well with the rest of the neighbourhood. What lightens things up a bit is the welcoming foyer with a small library, current exhibitions and a church guide with coffee thermos. The church is open during the day for all visitors. Notre-Dame de Pentecôte aims to be a church that welcomes all visitors to La Défense, regardless of their religious beliefs, and is therefore not primarily a parish church. It chooses to call itself a "Maison d'Église", i.e. a kind of "church home" that can serve as a respite for those who work in the neighbourhood and, one may assume,

3 On Notre-Dame de Pentecôte, see "La Maison d'Église de la Défense" (2025).

need a refuge from their stressful working lives. The term "maison" or "maisonée" is intended to evoke a "home" or "household" and thus provides a contrast or at least a counterpoint to La Défense as a workplace. It offers meeting rooms, lectures on various subjects, concerts, a communal lunch once a week and breakfast twice, and, of course, mass every day. Alongside the skyscrapers, the stylised and slender church tower can give an almost diminutive impression, but also helps the church blend in with its surroundings.

One of the most striking sculptures in the area is César's (1921–1998) bronze thumb, "Le pouce", at Place Carpeaux, a few hundred metres north-east of the Grande Arche. The thumb is César's signature; he sculpted thumbs since 1965. The La Défense thumb is his largest and tallest, measuring twelve metres. Place Carpeaux is primarily a transit zone or urban corridor, but in 2024 work is underway to transform the space into a more hospitable place, where there is a reason to linger for reasons other than looking at César's thumb or waiting for a taxi. The choice of a giant bronze thumb seems congenial to La Défense's monumental idiom, but can also be seen as an attempt to humanise it; the thumb is, as we know, essential to the grasping function of the human hand, and giving a thumbs up is an almost universal sign of approval. However, if you look at the thumb from a distance, passing humans appear very small. Unlike Barria's Art Nouveau sculpture, which opened this description of La Défense and can be found a few hundred metres away, César's thumb belongs more to a neo-realist, American pop art tradition. Here, too, one can read a tendency in the entire neighbourhood of La Défense to orientate itself not only towards the present but above all towards what it perceives to be the future: thumbs up for the monumental and progressive ambitions and self-images of Paris and France.

How well can a person settle in this neighbourhood? Can La Défense become a place for someone in the sense that it becomes existentially meaningful, a place of memories, existential foothold, community, good atmosphere and hopes? Or will both the spatial monumentality – in relation to the Grande Arche and the skyscrapers – and the temporal monumentality – in relation to the history of France – always overwhelm the visitor, so that the experience of La

Défense becomes one of muteness or, to use the term that will guide this chapter, placelessness? For those who like to lose themselves in the narrow streets of the Odéon or Saint-Germain or the classic arcades of the 2nd arrondissement, La Défense offers a completely different, almost hypermodern atmosphere. A Parisian I speak to in the city centre says he almost never goes there.

If it's true that belonging to a place means having an existential foothold there, what about all those times when we don't feel "at home" in the broadest sense of the word?[4] When we move to a new home, a new neighbourhood, a new city, a new country, because we have changed jobs, met a partner, escaped oppression, and don't quite feel at home, are we not in a place? We probably are, but we are in someone else's place, we are in this place as strangers. To be a stranger or a refugee is perhaps to have your physical "here" in a place that feels like "there". The process of finding a foothold in a new, unfamiliar place can be long or short, easy or arduous, but in the best case it results in us appropriating this place as a place where we, in some sense, "belong".

But aren't there times when belonging to a place never materialises? We can easily imagine, I think, the flight to another country where we are never really welcome. And what about experiences of a general lack of a spatially conditioned existential foothold, rather than a more specific alienation? Or with various cases of claustrophobia, agoraphobia or even topophobia?[5]

The rootlessness that both Hannah Arendt and Simone Weil wrote about was a rootlessness that was not a specific rootlessness, although it could be conditioned by a specific geographical displacement or by the fundamental destruction of one's home by the Second World War. Rootlessness had become generalised through the loss of a foothold. And this loss, as the metaphor of "rootlessness" suggests, is not independent of belonging to a place, as if it were an existential dimension alongside and independent of the spatial. The experience of rootlessness later came to be referred to in this expanded sense

4 Norberg-Schulz 1980, p. 23.
5 See Trigg 2017.

as just *placelessness*, for example by the phenomenologist Edward C. Relph, whose acclaimed book from 1976 is called *Place and Placelessness*.[6] However, we are always somewhere. So what could the meaning of placelessness be?

The Ability to Orient Oneself

In his 1960 book *The Image of the City*, American architect and urban planner Kevin Lynch studied how residents of three US neighbourhoods – Boston, Jersey City and Los Angeles – moved around and perceived their place in the city.[7] A neighbourhood is a structure that Lynch analyses in terms of paths (streets, sidewalks), edges (walls, rivers, buildings), districts (internally connected parts of the city), nodes (gathering places, intersections) and landmarks (important buildings, mountains or signs that can be used as orientation points). However, these are not significant in isolation, but in relation to each other, in the way that together they structure a neighbourhood. The fundamental concept in Lynch's analysis is *imageability*, which is essentially the ability to create, in a figurative sense, an image – a "map" – of the neighbourhood that allows one to orientate oneself there with some sense of belonging and overview.

Being able to "structure and identify" a local environment with the help of "definite sensory cues from the external environment" – to be familiar with a district – is an essential ability for being able to inhabit a place in an existentially satisfying way.[8] In doing so, we establish a mental image or map which is physical as well as social and affective, and a good such map (which is certainly in large part an affective and intuitive map of the infra-ordinary of the place) helps those who possess it to "establish an harmonious relationship between himself and the world".[9] But the possibility of creating such a map is influenced by the surrounding built environment; it

6 Relph 1976.
7 Lynch 1960.
8 Lynch 1960, p. 3.
9 Lynch 1960, p. 4.

can be facilitated or impeded depending on how the environment is constructed. It is not that the environment determines how we orientate ourselves. It is an interaction between human beings and environment: "The environment suggests distinctions and relations, and the observer – with great adaptibility and in the light of his own purposes – selects, organizes, and endows with meaning what he sees."[10] A good environment, according to Lynch, is one that is "open" to development through people's use of it and of which it is possible to create a mental representation. This may be more or less articulated, and in some cases it may be more accurate to speak of a bodily habitus rather than a representation. It is simply a matter of learning to recognise a specific environment so that it becomes possible to orientate oneself in it.

As Fredric Jameson has pointed out, this is not a map in the literal sense, and perhaps the term "itinerary" would be more appropriate, as it captures the dynamic nature of people's orientation skills.[11] However, both the "map" and the "itinerary" aid one's ability to orientate oneself in space using qualitative distinctions. Lynch lists a variety of characteristics that define the type of environment that lends itself to the existential need to create an image. These include aesthetic characteristics, but in his study identity and structure emerge as the most important elements. The distinctiveness of the physical environment – that it is distinguishable from other places – and the fact that its different parts somehow relate to each other, together with the meaningfulness it possesses for its inhabitants, appear as overriding characteristics for the possibility of creating a functional image, map or itinerary. That "the environment is visibly

10 Lynch 1960, p. 6.

11 Jameson 1991, pp. 51 f. Michel de Certeau points out on p. 120 (1988) that it was not until the 15th to 17th centuries that the map freed itself from the itinerary that otherwise constituted its condition of possibility. On the map as a condition of possibility for an overview that also concretely affects the understanding of place, see Bergwik 2024, especially pp. 282–291.

organized and sharply identified" means that "it will become a true *place*, remarkable and unmistakable".[12]

Again and again – as we have already discussed – phenomenology emphasises that spatial distinctions such as forward/backward, up/down, left/right are not just directional indications in a homogeneous space but qualitative distinctions.[13] While there may of course be cultural differences in the way these distinctions are understood, everyday language is saturated with such qualitative distinctions that belong to a symbolic order by which we orientate ourselves, existentially, in life. Reaching a clearer insight is like ascending a hill, and therefore a concrete ascent can also be understood as a quest for clear insight. That left and right have had a political meaning since the French Revolution is generally self-evident. Since the dawn of modernity, we Westerners, individually and collectively, strive forward, we want to make progress, but sometimes we also have to go back a little to understand where we went wrong.

Above all, such qualitative distinctions create the possibility of some kind of existential orientation in precisely the same way as Lynch's "imageability"; the totally qualitatively empty space would not only be a space in which it is impossible to orientate oneself, but also meaningless to do so. In a space without quality, there would be no order, no near or far, no here or there.[14] Or as Johan Asplund puts it: placelessness corresponds to "a mute world that can neither be known nor recognised".[15] It is therefore not surprising that Lynch's concept was incorporated into phenomenology and political theory fairly immediately. Norberg-Schulz refers to him in a discussion of the loss of place as a consequence of homogenising tendencies in modern architecture, while Jameson sees the potential of a more ideologically critical use of Lynch's critique of the alienated city.[16]

---

12 Lynch 1960, p. 92.
13 Cf. e.g. Bollnow 2011, pp. 18 f. *et passim*.
14 See Malpas 2006, p. 135.
15 Asplund 2022, p. 247. See p. 246.
16 Norberg-Schulz 1980, pp. 190 f.; Jameson 1991, pp. 51–52, 415–417. Jameson refers in this context to Louis Althusser's classic article "Ideology and Ideological State Apparatuses" (Althusser 2014).

In his investigation of placelessness, the aforementioned philosopher Relph takes the lead in an argument in which he speaks of significant places in a way that reminds us of Lynch's "imageability" and the need for qualitative distinctions. To be human is, according to Relph, "to live in a world that is filled with significant places", and to have one's *own* place in this world.[17] Places are characterised precisely by qualitative and intersubjectively shared distinctions that visitors and residents experience in a specific place and which they themselves help to create, albeit not in the same deliberate way as a particular architecture. Relph emphasises how people are co-creators of places, with one of the most important acts being naming. Naming something means, we could say, at the same time distinguishing a place from all other places (Lynch's identity) and putting it in relation to them by indicating what this particular place is (Lynch's structure). But it is not only about participation in the ongoing use of a place, but also about being recognised.[18] Relph, like many other phenomenologists, likes to emphasise active participation as a way in which the place is constituted as a place. Still, he acknowledges there is a mediopassive dimension.

To be recognised, to be seen, implies a recognition that I cannot give myself, but which comes from an existential belonging to a place. Sitting in his café in Place Saint-Sulpice, Perec sees a number of familiar faces pass by, occasionally greets one of them, and generally his little book gives the impression that he belongs there. To be a stranger is in some sense to lack such existential belonging and therefore not to be recognised. Interpelled, perhaps, but not greeted by more or less fleeting acquaintances with a recognisable nod. Furthermore, it can also be said that a significant place is a place where I can feel bored, because the place becomes too familiar through habit. The experience of place is a dialectic between homesickness (*Heimweh*) and *Wanderlust*, and this dialectic only becomes concrete in relation to a significant place.

17 Relph 1976, p. 1.
18 Relph 1976, p. 37.

A loss of these qualitative distinctions in the experience of places follows from placelessness, as modern buildings, neighbourhoods and districts become more and more similar to each other, and this also means a kind of namelessness, which goes hand in hand with the absence of qualitative distinctions to orientate oneself. The experience of the environment becomes "chaotic, lacking in orientation, even fearful", as it has no "humanised and familiar points of reference".[19] All directions, all surfaces and spaces are experienced as equivalent, and their existential meaning is thus lost. Or the environment becomes flat, superficial, without "intentional depth".[20]

The placelessness here is not primarily an effect of being a stranger in a particular place, an "existential outsider" who does not belong to the place, who does not feel at home and who has difficulty orientating himself. It does not matter whether this is a self-chosen, deliberate strategy or an unintended consequence of a forced flight or displacement. Not identifying with a *particular* place may indeed be experienced as homelessness or rootlessness in relation to *that* particular place, but there is a fundamental point in distinguishing this experience from a more all-encompassing alienation. Nor is it a matter of *some* places being perceived as lacking their own identity. Such places have existed in all societies. What is new is the scale with which this is enforced in modern times. Relph refers to the French sociologist and theologian Jacques Ellul and his concept of *la technique*, which Relph understands as a quest for "functional efficiency, objective organization, and manipulative planning" that takes no account of local conditions.[21] It is based on an understanding of space as uniform and homogeneous, simply as a resource that can be manipulated at will, and where qualitative distinctions of an existential nature do not matter. The purpose is big business, tourism, mass culture, political authority. For Ellul, what characterises *la technique* is above all that it is the smoothest possible way of achieving certain predetermined goals, but where

19 Relph 1976, p. 17.
20 Relph 1976, p. 79.
21 Relph 1976, p. 81.

efficiency has become an end in itself.[22] In other words, there is an instrumentalisation of space, where space is reduced to mere locality. The space being planned is removed from the spaces where we live and experience our everyday lives. It no longer holds any existential distinctions.

The French anthropologist Marc Augé – with some influence from Michel de Certeau – has coined the term "non-place".[23] Augé calls "anthropological places" those places that are constituted by identity, relationship and history, i.e. those places that, according to Lynch, are possible to form an image of.[24] Non-places are thus places that are less socially and symbolically charged than these. Non-places are a product of "supermodernity", a state of abundance in late capitalist society characterised by the "shrinking of the planet" in terms of spatiality. We can reach virtually any place by modern means of communication and they reach us through television, the Internet and so on. Paradoxically, this isolates the individual from others, as another effect is that it becomes more difficult to establish spatial community. There are, to allude to Arendt, no tables to gather around. The typical spaces of supermodernity are spaces where we are indeed together with other people, but as solitary and anonymous individuals: "As anthropological places create the organically social, so non-places create solitary contractuality."[25] Typical non-places, according to Augé, are motorways, airports, supermarkets, hotel chains, shanty towns, refugee camps, which are designed for us to pass through as smoothly as possible, without us leaving any personal traces or them taking on any special meaning for us. The non-place is characterised by its uniformity; it corresponds to a global world without any thresholds where people compete to build the tallest building in the world. According to Augé, places and non-

22 See Ellul 1964. Nolen Gertz brings Ellul's critique into our time from a Nietzschean perspective in Gertz 2024.
23 Augé 1997. See de Certeau 1988.
24 Augé 1997, p. 42.
25 Augé 1997, p. 76.

places coexist even in our contemporary world, but it is above all the non-places that characterise the supermodern world.

It is easy to understand both Augé and Relph as nostalgic critics of placelessness, since the former speaks of place in terms of an organic sociality and the latter distinguishes between authentic and non-authentic places. However, Relph has two qualifications of the concept that are important to note. First, he emphasises that the experience of meaning in places is not dependent on them being particularly remarkable places or built in a particular way; people have an ability to make themselves at home – to "put down roots" – experience meaning and care about the places they inhabit, as long as there is the possibility to make qualitative distinctions, to be in some sense part of and even have a name for these places.[26] One does not have to pass the Pantheon, the coast of the Western Cape or see the Manhattan skyline every day on the way to work to realise that one lives in a significant place. Even these extraordinary places can eventually be transformed into something ordinary by habit. Our everyday lives do not have to be lined with spectacular places, but they must be identifiable and orientable. The place does not determine the person, but is created in an interaction between place and person.

Secondly, Relph does not believe that the places that are important to us must have a stable and unchanging continuity over time. On the contrary, places change all the time, and necessarily so, but the change still leaves room for a continuity that enables an existential attachment to them. It is true that places can die by being destroyed, abandoned or no longer used for their original purpose. But many places survive through change, even fairly radical change. Process rather than thing, then. The fact that places change is thus an inherent condition for their survival. Paris could be an example. Despite all its radical changes, its history is sedimented not only in memories, but also physically in the placement of houses, streets and squares that tend to survive over centuries, not necessarily as concrete constructions but as structures. This is particularly true of public

26 Relph 1976, p. 18.

places with high imageability – landmarks such as Red Square in Moscow, Niagara Falls and the Acropolis, Relph's examples – which have attracted attention under many different political regimes.[27] Artificially preserving history can easily become an idealisation of it, like living in a museum.[28] The place must avoid both a nostalgic idealisation of history and leave room for human participation in order to become not a static but a dynamic phenomenon.

The characteristics of placelessness according to Relph are: uniformity, lack of human scale and order in buildings and places, mass production, organisation dictated by the economic interests of large companies and the political administration, an attitude characterised by *la technique*, kitsch and so on.[29] How accurate are these points? La Défense in Paris will serve as a test bed.

The Saturated Place

It is perhaps obvious from my description of La Défense at the beginning of this chapter that I regard it, on the whole, as a place, if not a particularly hospitable place, at least not entirely placeless. But such a statement needs to be developed: placelessness is, after all, not an absolute and unambiguous phenomenon.[30] One way to approach the experience of La Défense is as an experience of something overwhelming. The question then becomes in what way the neighbourhood overwhelms us.

The experience of the existentially significant place is an experience of what the French philosopher Jean-Luc Marion calls a *saturated* phenomenon.[31] The overwhelmingness of these saturated phenomena is expressed by Marion as the fact that intuition receives

27 Relph 1976, p. 35.
28 Relph 1976, p. 101.
29 Relph 1976, pp. 118–121.
30 According to Siegert 2024, pp. 14–16 *et passim*, it is rather *the sea* that is characterised by a pure placelessness than anything on land. I shall return to this elsewhere.
31 See Marion 2002a, §23.

more than what intentionality aims at: it is not the direction of my consciousness towards the phenomena that is primary, but the way in which they impose themselves on us. *Poor* phenomena, Marion says, are those that we cannot experience at all, such as mathematical and logical intuitions. With *ordinary* phenomena, we know, so to speak, what they are before we even encounter them, a kind of functionalism where we master with our minds what they should be and where the phenomenon becomes an object or product, if you like. When we encounter the product, it only confirms the idea. My coffee maker in the office lives up to what I expect from a coffee maker; it is, as Marion puts it, an "anecdotal appendix" to the idea of a coffee maker, because it adds nothing new. Only *saturated* phenomena overwhelm the recipient, when "intuition sets forth a surplus that the concept cannot organize, therefore that the intention cannot foresee".[32] Thus the order between phenomenon and recipient – for example, between place and person – is reversed in relation to the ordinary phenomena; it is the phenomenon that presents itself to the recipient before and independently of the recipient's purpose for it. In this case, we cannot anticipate what is to come. In fact, this is a fundamental mode of understanding our relationship with the world. The world presents itself to us, Marion argues, as a saturated phenomenon; our claims, purposes and hopes for and with the world come second.

There are four – or five – types of saturated phenomena: event, idol, flesh (*la chair*) and icon – and also revelation with a small "r". The first four saturated phenomena correspond to Immanuel Kant's four categories of understanding: quantity, quality, relation and modality. The event refers to the unexpected, that which cannot be comprehended or described in an exhaustive way but which, for example, has a decisive impact on history, such as the Battle of Waterloo. The idol denotes the visible phenomenon that captures the eye and holds it captive, such as the sublime work of art. The flesh is about the identity between what is affected and what affects in a *fait accompli*, as in the case of joy, love, suffering, anguish, and so

32 Marion 2002a, p. 225.

on. The icon, finally, allows the invisible, that which withdraws and can never be reduced to my own self, to appear: the face of the other. The fact that the icon can never be reduced to the gaze with which I look at it is because the icon has the initiative. It is the icon that casts its own gaze on the one who looks at it; I experience that I am being given an experience, so to speak. It transforms the self into a witness. The fifth and last possibility according to Marion is what he calls revelation with a small "r", which does not add a new element, but varies the four phenomena, combines them and plays with their possibilities.[33]

The quite everyday, infra-ordinary experience of a particular place of existential significance is precisely a saturated phenomenon.[34] In the infra-ordinary experience of a place, intuition precedes intention. Or, in other words, my experience of the place as this particular place will precede all the ways I want or intend to use it and relate to it. It's not so much about the people and their places, but more the places and their people.[35]

I experience the place with my body. Sensation, taste, smell, hearing, sight and existence. But with the flesh in the autoaffective sense that Marion gives to the term, I experience myself together with the place. The experience of place is, probably almost inevitably, an experience of its infra-ordinary character, the essence of which is that it both precedes and is rarely registered by intention. Perec and his attempt to exhaust a place in Paris is thus an exception, since there he is precisely trying to uncover its infra-ordinary features. But if the infra-ordinary can be understood as those qualities, events and things we *hardly* notice about a place, then it is in the nature of things that the experience of the infra-ordinary in a place is a curiously anonymous experience. If the infra-ordinary is a saturated phenomenon, it is therefore a phenomenon that envelops us even

33 Marion 2002a, §24.

34 Marion often appears in his examples as oculocentric, so this claim implies some criticism of his phenomenological project.

35 Cf. Marcus 2023, p. 140.

when it is not explicitly the object of our attention, like and in the form of an atmosphere.[36]

But Perec failed, as I said, in his attempt to exhaust a place in Paris.[37] First, because Place Saint-Sulpice is inexhaustible in its phenomenological richness – in principle, like any other place – and therefore simply cannot be fully described, even by the most skilful writer; secondly, because the uniqueness of the place is lost in the description of all those everyday trivial events that Perec records. As the literary scholar Robert T. Tally rightly notes, bus 86 could just as easily be the bus passing a café in Durham, North Carolina, as the one passing the Café de la Mairie. But what Tally fails to mention is that in Perec's account it is not the detail of bus 86 stopping outside the café that is his infra-ordinary point, but this detail in combination with all the other "micro-events" and details such as trees, pigeons, vehicles, people, the restaurant's salad, and so on that he accounts for. Perec doesn't so much try to isolate unnoticed individual details of everyday life, as to emphasise the wealth of trivial details that belong together precisely by virtue of belonging to a concrete place. He does this in order to allow the infra-ordinary texture of the place to emerge from its relative oblivion for a moment. Sure, there may be a bus 86 in Durham and a hundred other places. But if a place in Durham had been the subject of an attempt at exhaustive description, the almost equally inexhaustible configuration of infra-ordinary details that can be found there would have given its unique character to that particular description.

There is another important difference in existence between the person who has the infra-ordinary experience of the café at Place Saint-Sulpice on the spot and the person who merely reads about it in Perec's text. The perfectly correct statement that Perec fails in his intention to exhaust a place in Paris says something, not primarily about Perec's ambition or ability, but about the *intention* as such. Daring to attempt to *exhaust* something in writing, however

36 This is reminiscent of what Marion calls a "banal" saturated phenomenon in Marion 2008, pp. 119–144.

37 Tally 2013, pp. 53 f.

delicate, flexible and skilful, is an example of treating what Marion calls an everyday phenomenon, i.e. a phenomenon that we try to master with our minds, as when we wonder whether the Café de la Maire corresponds to our idea of a classic Parisian café. When Perec describes Place Saint-Sulpice, he does so mainly with the aid of sight, so Søndergaard's attempt to eavesdrop on the same place will complement rather than replace Perec's attempt. But if the concretely given experience of all the infra-ordinary qualities of Place Saint-Sulpice is in fact synaesthetic and multimodal, then any such attempt, whether it be one that emphasises sight or hearing or any other sense, involves a reduction.[38] The more precise meaning of Perec's failure, therefore, is that Perec's intention to try to exhaust a place in Paris can never live up to the intuition of its infra-ordinary constitution. Perec's experience of Place Saint-Sulpice and its inexhaustibility, sitting in one of its cafés, overwhelms him. Intention can never live up to intuition. Perhaps the failure is in fact intentional, and thus in a sense not a failure at all. It is about allowing an experience to rise to the surface that can never be addressed directly.

In Marion's terminology, the infra-ordinary is therefore more reminiscent of the flesh and its invisibility than of the idol with its visibility.[39] The infra-ordinary of the place that makes itself felt synaesthetically does not overwhelm with its visibility, but in an invisible way, because it is more fundamental, more intuitively given than the intentional relationships established with the things around us. If the flesh constitutes the identity between the affected and that which affects, it means that the flesh is both the experience of myself as a corporeal being who is really *there* on the café chair and the condition for the world of phenomena to open up to me. The (relative) invisibility or anonymity of flesh is found amidst all its visibility, audibility and general possibility of sensation, and in this it is like the icon. But its appeal belongs to the everyday or ordinary

38 Cf. Sigurdson 2024c, pp. 39 f.
39 For a more detailed discussion of flesh in Marion, see Marion 2002b, pp. 82–103. For a broader background to the history of the concept, see Dahlberg 2013.

rather than the extraordinary; it does not *want* anything from us in the same way as the idol or icon. In a similar way to place, flesh is ultimately the medium in which micro-events take place and in which we establish relationships with things, rather than the events or relationships themselves.

Places give themselves to the person who is there with a dimension of depth that comes from somewhere other than one's own consciousness, and they assert themselves as an abundance towards the consciousness that receives the phenomena of a place. But they do so with such obviousness that it is easy to take them for granted and reduce their importance for our existence. We mistake them for what Marion calls a *common* phenomenon. Calling them the background rather than the focus of all that we do and see in a particular place captures something accurate and essential, but only if we do not thereby think we can be independent of them or treat them as an objectified scene of our activities. They are background in the sense that they are an ever-present, preconceptual condition for all our articulated or thematised insights and claims about places.

Undeniably, the tendency to see places only from the perspective that we create them, thus turning them into what Marion calls a common phenomenon where intention trumps intuition, is not uncommon. But nevertheless, the tendency is illusory insofar as it leads us to consider ourselves as those who construct and rule over places, rather than those who in various ways inhabit, visit or reside in them, or in other words fulfil the role of witnesses to places. Places are the tacit condition of our existence. The knowledge we undeniably have of them – and which we must have in order to orientate ourselves there – is a tacit knowledge in Michael Polanyi's sense.[40] There is therefore also an irreducible asymmetry in them; neither Perec, Søndergaard nor anyone else can really "exhaust" a place in Paris. The sensory excess of the place exceeds what receiving subjects – me, sitting in the café at Saint-Sulpice – are able to receive, but also what they are able to grasp, articulate and understand.

40 Polanyi 2009.

Can Marion's description of saturated phenomena help us to articulate the problem of placelessness? According to this way of approaching the problem, an absolute placelessness must be a limit concept. Whatever we think of non-places, whether we think of a specific house, a neighbourhood, a city or a whole region, they cannot be without all kinds of infra-ordinary qualities, any more than they can be completely devoid of atmosphere or resonance even when, for the visitor, they approach the limit of muteness. Rather, it is about the *way* the non-places embody these qualities.

Phallic Verticality

One way of understanding a neighbourhood like La Défense is that it has a sensory excess that approaches that of the idol, thanks to its visibility. There is a verticality here that overwhelms the visitor: the skyscrapers, the arch, but also the church tower and the thumb. Not all verticality has to be phallic; the Gothic spires, buttresses and pointed arches reach upwards in a quest for something else, born of a realisation of our own human finitude.[41] But in the case of La Défense, it is hardly wrong to speak of a phallic symbolism that expresses ambition, expansion, control and power.[42] There may indeed be a temptation to interpret a kind of functionalism in a neighbourhood such as this, a functionalism in which, in accordance with Marion's characterisation of ordinary phenomena, the concept precedes the thing. But like many functionalist constructions, they embody a vision of human beings that transcends mere function. In the case of La Défense, we cannot ignore the monumentality and the vision of France that is expressed here in this phallic verticality. It is a question of the ability to survey and control, of a visible height rather than a dimension of depth. Because of the overwhelmingness of height, the impression of what the architectural historian Dolores Hayden speaks of as "procreant power" and "awe", visibility comes

41 See Ruskin 2009; Simson 1989.

42 Certeau 1988, pp. 91–93; Hayden 1977, pp. 108–115; Lefebvre 1991, pp. 286–290.

to dominate the other senses, to the point that the infra-ordinary, synaesthetic and silent texture is even more in danger of being hidden and forgotten.[43] Like the idol, La Défense captures my gaze, and thus not only my gaze but my entire existence.

Faced with such verticality, I am almost haunted by a need to escape, by going into one of the many restaurants, the shopping centre or perhaps the church *maison*. They protect me not only from the sight of the skyscrapers, but perhaps above all from the way in which these interrogate me. The monumental architecture places me in a relationship where its immense volume emphasises my own smallness. Phallic architecture embodies potency, agency, endeavour, control. It is born, among other things, of a quest for functionality that either does not recognise its own vision as a vision, but understands it as a mere neutral instrument, or else considers its own vision to be the only rational one, in step with inevitable progress. It confuses a saturated phenomenon with an ordinary one, or perhaps even a poor one, as if there were a pure instrumentality without transcending intuitions, as if it were possible to turn human spatiality into a mere scenery that we are able to construct and therefore master. Such functionality is metaphysical in its intention, in Marion's sense, a domesticated hermeneutics, which means that it strongly insists on wanting to set the framework for how La Défense is to be encountered and experienced. There is an inherent tension here: by virtue of a vision of human agency and control, the visitor will be interpellated into a passive rather than participatory relationship with this place. They become a witness to the monumentality of La Défense, but not as a participatory witness. Instead, they are a witness whose role is merely to be amazed and impressed. The place will never be ours or mine. It has undeniable characteristics of non-place, at least if non-place and place are not perceived as mutually exclusive alternatives. It is largely unresponsive, not asking for my or our response; it overwhelms by intruding rather than inviting.[44]

43 Hayden 1977, p. 111.

44 Cf. Asplund 2022, p. 246: "A place is a structured physical space of some permanence, to which one can ask questions – and get answers. A place is

The phallic verticality of La Défense can be called placeless to the extent that it makes my practical participation in the neighbourhood difficult. I become its consumer not, through a participation caused by my response to it, its producer.

The Ambivalent Place

The placeless place is, we could say, the paradoxical attempt to build an abstract space in the concrete world of the senses. Henri Lefebvre, in his 1974 *The Production of Space*, like Relph and Augé, places abstract space in a historical period, the era of the bourgeoisie and capitalism, which he believes is linked to an exchange of goods and words.[45] It is even the case, according to Lefebvre, that abstract space is based on a (supposed) consensus that tries to subordinate all other spaces. Abstract space is an ideational representation of spatiality that claims hegemony, but which is nevertheless only a conception of space and as such not identical with lived space, even when, as an ideal characterised by *la technique*, it comes to shape the way concrete space is constructed in buildings, streets, squares, monuments and so on. Undoubtedly, concrete space influences how we live our lives, the representation of space and space as representation can mutually support each other, as when the monumentality of La Défense also gives an impression of efficiency and progress well in line with the economic training and activities undertaken in the district. But the shaping of concrete reality is rarely as frictionless as its representation suggests; the cafés, the artworks, the church of La Défense are potential points of friction, ready to be put to use in a way that does not – at least not immediately – support the atmosphere of monumental *technique* that the neighbourhood embodies. "Abstract space *is not* homogeneous; it simply *has* homogeneity as its goal, its orientation, its 'lens'. And, indeed, it renders homogeneous. But in itself it is multiform", writes Lefebvre.[46] And this enables a use that

'alive' in the sense that it *responds*."

45 Lefebvre 1991, p. 57.

46 Lefebvre 1991, p. 287.

runs counter to the notion of unlimited efficiency that characterises the representation of abstract space.

Michel de Certeau, in his influential book *The Practice of Everyday Life*, has drawn attention to the way in which the everyday user of a place, a thing, or a statement cannot avoid being determined by the intention of the producer or the structure of this entity. He illustrates his point by comparing the view from the World Trade Centre, which still existed at the time, with the experience of walking in New York City. The view from the skyscraper here becomes a symbol of the will to survey, the active attempt to shape and control a space. It becomes the epitome of a "functionalist totalitarianism" that seeks to counteract any local cultures, since these "compromise the univocity of the system".[47] But because it suspends the place as a symbolic system, it implies an "annulment" of the "habitable city".[48]

Walking in the city, however, allows no such overview. Here one does not so much act as react. The circumstances – buildings, streets, car drivers, cyclists, other pedestrians – have to be dealt with at the very moment they appear. It doesn't work to stick to a plan, because no such overview is possible when moving at street level. At any moment a cyclist can appear and I, on foot, have to give way.

This is the difference between a strategy and a tactic, according to de Certeau: a strategy can only be attempted by those who consider themselves to have an overview; the rest of us have to take it as it comes. This is also how de Certeau distinguishes between different kinds of power: the power to decide and shape on the one hand, and the power to make do with what you have on the other. Many everyday practices, such as speaking, reading, walking, shopping, cooking and so on, are tactical rather than strategic in nature. They are a kind of rhetoric, in the sense that they manipulate a system; they make use of it in a way that contradicts any claim that it constitutes

---

47 de Certeau 2002, p. 106. It should be noted that de Certeau uses the terms space and place in a different way than I do here, see p. 117: "*space is a practiced place*. Thus the street geometrically defined by urban planning is transformed into a space by walkers".

48 de Certeau 2002, p. 106.

an established order and is therefore final, whether it is a language or a place.

The philosopher Dylan Trigg has warned that the dualism between place and non-place has become too antithetical in Augé and Relph, as if the heterogeneous historical and relational nature of place could be easily contrasted with the homogeneous, historyless and non-relational nature of non-place. While Augé and others have a point about the homogenising tendencies of the modern, there is a risk in this, namely that the ambivalence of place as such remains hidden. Places cannot be divided in any simple way into "home" or "away", "authentic" or "inauthentic" (and in the latter case non-places). If this were the case, place would be a finished product rather than something in which we can participate. For Trigg, "place is such that its formation is a relational movement, in which the interaction of self gives form and plenitude to place from the *inside out*"; place is "a source of strangeness rather than being determined in advance".[49] Whether a place is experienced as a place or as a non-place has, according to Trigg, to do with repetition and habit rather than objective characteristics of the place as such.[50]

I would like to add an "only" and an "also": the experience of a place as a place has to do not *only* with objective characteristics, but *also* with repetition and habit. What would happen if I lived in La Défense? Wouldn't some parts of the neighbourhood eventually take on a more, for me, familiar and even participatory character, in line with what de Certeau says about the ability of everyday practitioners to make do with what they have? Would it be impossible in principle to establish an existential foothold here? I am not asking about the attractiveness of such a proposal, but only about the possibility. Furthermore, is there not always the risk of projecting our own personal and cultural sympathies and antipathies onto buildings, environments and urban plans that are in some sense alien to us? My relationship to La Défense is undeniably determined by the places

49 Trigg 2012, p. 126.
50 Trigg 2012, p. 132.

that have shaped me in the past, the habitus built up by my habitats, and the preferences that I have made my own.

However, Trigg's critical remark goes beyond this, in pointing out that the place can never be reduced to something familiar, homely, accustomed – again an attempt to domesticate places, to make them lose their overwhelming qualities. Nor for Martin Heidegger, in his classic essay "Building Dwelling Thinking", does the human act of inhabiting a place reduce it to a static domain that can be mastered.[51] There is something uncertain, ambiguous and even uncanny in a place as such, something we cannot capture but also cannot avoid by relocating it to another place that we call, for example, the "non-place". All talk of placelessness easily becomes such an endeavour, which is why Trigg prefers to use the negation prefix *un*, as in "un-place", rather than "non-place". The prefix signals, in this context, a contradictory tension *with* the place rather than an alternative *to* the place, like the "un-dead" zombies who are neither dead nor alive. The strangeness of place here is clearly not just an absence of all the character that characterises a place, but the possible presence of a character that can never be domesticated by human projects. Here, Trigg is approaching the argument I made in the previous chapter about the spirit of place.

This is not entirely unknown to Relph and Augé. The latter writes: "The possibility of non-place is never absent from any place."[52] In other words, places and non-places are intertwined. At the same time, as Trigg points out, there are risks in contrasting them, especially if, like Relph, one also speaks in terms of an authentic sense of place in the sense of belonging to a place as an individual and as a social being.[53] Apart from the risk of domesticating place by attributing to it overly familiar qualities, as well as hiding all the intermediate forms that exist between place and non-place – not least the non-place is, due to the human bodily constitution, a limit concept rather

51 Heidegger 1975. See Malpas 2012, p. 154 and *passim*; Malpas 2006, *passim*.
52 Augé 1995, p. 86.
53 Relph 1976, p. 65.

than an actual possibility – there is also a sense in which we and every society really need the non-place in its non-relational and impersonal sense: as a common space where we bracket the personal (as far as possible) in order to negotiate with each other.

This concerns the table I mentioned apropos Hannah Arendt in a previous chapter, the table that both unites and divides those who sit around it. It is about a public space that is neither home nor away. It is also about Kungsportsplatsen, which certainly carries its history, its identity and its structure, and thus really is a place, but which is designed for people to be transported and interact with each other in a way that does not create any deeper relationships, and in this way also carries something of the non-place in itself, but without its more dystopian aspects. Here, place and non-place are tangibly intertwined, albeit in a less antagonistic way. Some days Kungsportsplatsen is a place where, like Perec in Place Saint-Sulpice, I meet acquaintances and let myself be captivated by the surroundings; on other days I pass through as a mere transport on my way somewhere else. As Lars Marcus rightly points out: "Places of anonymity have their place in larger cities".[54] Multifunctional places borrow, one might say, a feature of the non-place precisely in that they leave room for a use that is more instrumental, without this necessarily obliterating their genuine place character. In the modern city's endeavour to reduce human suffering there is both something desirable and something utopian, i.e. something literally placeless.

There is no reason to deny that many people in our time experience a genuine sense of alienation from place, that there are homogenising tendencies, which risk reducing the ability to orientate oneself in relation to place, or that indiscriminate demolitions wipe out memories of place. In this sense, there may be reason to talk about placelessness, if not in an absolute sense, then at least in the sense of Johan Asplund: "Placelessness is [...] *impersonal*."[55] Placelessness does not really mean a loss of place as such, but a supposed understanding of place as nothing more

54 Marcus 2023, p. 142.
55 Asplund 2022, p. 248.

than geographical locality. As Jeff Malpas writes: "Even under the reign of technological modernity, our relatedness to place is not obliterated, but is rather covered over, ignored, made invisible."[56] Placelessness as an experience of lacking an existential foothold is probably related to the relative absence of resonance – of a place-bound response that is not merely a projection of our imagination but rooted in history, buildings, the structure of the neighbourhood, its social composition and so on.

However, there is a need to talk about place that does not succumb to a dualistic distinction between home and away. There are existentially significant places that are neither home nor away, and there are places that are away rather than home but are still significant. Rather than a placeless place, La Défense is a partly inhuman place, and in this very way significant. A more comprehensive treatment of existential significance must include a more general discussion of how places charged with it relate to each other in different ways. One aspect of this concerns topophilia, and in particular a concept and experience that together carry a double ambivalence: feeling at home. Let us now turn to this complex concept that has haunted this whole investigation of place.

56 Malpas 2012, p. 63.

Home is a complicated concept. Its complications have to do, firstly, with our concrete experience of home. There are all those times when we are unhappy at home, when home may seem like a place without space, where we cannot breathe. Or there are those cases when home has lost its dynamism, when it becomes so familiar that being at home results in boredom, or where the home becomes the scene of a traumatic event and takes on an eerie atmosphere, or where the home has become occupied, inaccessible or erased. For many, the home is the most dangerous place, as it is where domestic violence is most common. Given the fundamental importance of the home, both practical and emotional, a disruption or breach of the sense of home will have major consequences for the individual as well as for the social community. Not being at home in one's body, nor in a place one considers to be one's own, nor in a shared world, constitute acute forms of existential alienation.

The complications of the concept also include all the different senses in which we use it. We speak of "homeland", of "father-" or "motherland", but also of "home" in the sense of dwelling. I may "feel at home" when I pass through Kungsportsplatsen again after a period of absence, but that does not mean that my habitat is in Kungsportsplatsen or that I would like it to be. It seems that a fundamental meaning of "home" has more to do with "dwelling"; the French *maison* and the Italian *casa* are more obvious here in that they combine the physical aspect of home – an apartment or house or other building – with the affect that also characterises the experience of home.[1] In Italian, no-one would think of calling Kungsportsplatsen *la mia casa*, but that does not mean that the experience of being in Kungsportsplatsen is an experience of being away. In everyday language, this is probably not a big problem, but when you approach home and away as concepts and phenomena, it turns out that these concepts offer a rather rough division of the experiences we have of places. None of the places I have used in this book as examples of places where it is possible to feel at home – my office, Kungsportsplatsen, Place Saint-Sulpice – are for me a home, in the sense that *la mia casa* is home. It is important to realise that our interaction with the places we frequent is experientially much more nuanced than the home/away pair manages to portray, not only to do justice to the experiences but also to avoid exaggerated or misguided expectations of many of our existentially significant places. There are alternatives to the workplace being perceived either as impersonal or as an extension of *mia casa*.

That I can claim in everyday language that I feel at home when I pass Kungsportsplatsen need not be a mistake. What I am expressing is, among other things, that existentially significant places are charged with a certain kind of affect. To "feel at home" in this figurative sense is to harbour a sense of belonging to a particular, concrete place. This is a place that in some sense not only belongs to me, but to which I belong, more than other places. "Feeling at home" is precisely an expression of affect, of what is usually called "topophilia". Topophilia literally means "love of" or "friendship with place" and is defined by Gaston Bachelard as "felicitous space": the kind of space that we cherish, protect, are attracted to, love, in short, "eulogized space".[2] The existence of uncanny homes, of inhospitable homes, of homes in which we are not given any space, does not contradict this existentially fundamental affectivity with which we inhabit spaces, but rather emphasises it, albeit *ex negativo*.

1 See Coccia 2021.
2 Bachelard 1994, p. xxxv.

A final complication is the dialectic of home between gift and task.[3] In Homer's *Odyssey*, the home on Ithaca is not only a geographical but also an existential point of orientation for both Penelope and Odysseus. Home is the place that makes them who they are.[4]

Penelope's experience of home cannot be described as merely static, in contrast to Odysseus' dynamic adventure on his way home from Troy. The permanence of a home is produced by repetition, such as shopping for food, cleaning, maintaining water, heating and sanitation. For Penelope, repetition is symbolised by her work at the loom while waiting for Odysseus to return home. Weaving is a repetitive activity. However, when Odysseus' return is delayed, there is a suspicion that he may never return. Suitors flock to her. When the weaving is finished, she promises, she will choose among them and marry. But in secret, every night she unravels what she has woven during the day. In doing so, Penelope extends her deadline, creating another space in her home. By introducing a difference – through a literal unravelling – her experience of home changes from something she merely reproduces to something she actively produces. Her experience of home is now of both repetition and difference, of home as both starting point and aspiration.

There is a distinct difference in the quality of the experience of being here at home and being over there. Penelope is home, Odysseus is away. But the asymmetry of this relationship – that away is defined in terms of home rather than vice versa – need not be understood as contrastive. That would be to ignore the fact that both home and away are dynamic phenomena: *over there* is not automatically transformed into at *home here* except in a strictly geometric sense by the body's *here* moving to the foreign place's *there*. More is needed to turn an *over there* into an at *home here*, namely Penelope's repetition in all its senses. Both *here* and *there* are certainly readily borrowed in nostalgic attempts to characterise

3 On this dialectic, see Sigurdson 2026.

4 Here I have drawn inspiration and insight from Sand 2008, pp. 11–14 *et passim*. Reflections on Penelope and Odysseus and their relationship to the home are interspersed like a red thread throughout this book.

home as the familiar, even the authentic, and away as the foreign and inauthentic. But that too would be to ignore the fact that even in their asymmetrical relationship, *here at home* is no less dependent on *over there*; if a home were entirely without the space of an *over there*, it would be like a prison and thus not a home at all. The very experience of place as something given, not something I created on my own, but at the same time preserved through repetition, prevents the place from being reduced to something too familiar. There is always a touch of absence even in what is most at home. When Odysseus returns home one day to Ithaca and to Penelope, he is no longer the same. For Penelope, too, home is different, and she has been forced by circumstances to make it so herself by changing her weaving technique.

As I wrote in an earlier chapter, our habitat is shaped by our habitus and vice versa, and it is no coincidence that these terms, habitat and habitus, are related to each other, like the English *habit* and *inhabit* or the German *Wohnung* and *Gewohnheit*. The act of inhabiting a dwelling in the sense of making oneself at home, for example when moving, is equivalent to the Greek term *oikeiosis*, which could be translated as "settling in". Settling in, so to speak, is not a creation out of nothing, but requires resonance and responsiveness to the place that will become my habitat, a kind of fundamental trust that it is possible to inhabit. As the philosopher and psychiatrist Thomas Fuchs points out, such an act presupposes "a storehouse of the sedimented experiences of others", experiences that are at once physical-embodied, cultural and social.[5]

Recall the quote from Annie Ernaux in the introductory chapter to the second part of this book, which describes how gestures, styles of walking, ways of speaking are transmitted from "body to body" between people "from the remotest countrysides of France and other parts of Europe".[6] The inter-corporeal, the cultural and the social is the way home is given to us. Even in our doubting, criticising or moving away from home – all of which are contained

5 Fuchs 2024, p. 130.
6 Ernaux 2017, p. 26.

in Penelope's ability to unravel the fabric she has woven and which is also the fabric she herself is woven from – we inevitably draw on this more original domesticity.[7] The fact that home is given to us as a gift precisely through its repetition means that there is always, in principle, a conditional space for rupture.

These complications mean that our experiences of what actually constitutes a home need not be the same. Not only the settler, but also the nomad has a home. Nor must the settled home be a single one. What is common is probably, as Fuchs puts it, an atmosphere of "homeliness, comfort and protection" with which we are familiar and which relates to concrete places.[8] And yet there is always the possibility of an experience of away in home.

Returning to this experience, I shall summarise the results of this study of places and their relationship with us who inhabit them. In keeping with its subject, this final chapter will therefore revisit places that are already familiar not only to the reader of the previous chapters, but to every place-bound human being: *loci communes*. We shall move from the home – *la maison* – and its importance for our anchoring and orientation in the world – navel, *umbilicus*, *ho omphalos* – through the sense of home in an expanded and controversial sense – *die Heimat* – to the threshold of something else.

La maison

First: the home and its importance for topophilia.

Bachelard's 1957 *The Poetics of Space* is a classic text for understanding topophilia, but it is more concerned with describing the phenomenon than defining the concept. In addition to the definition I have already quoted above, Bachelard explains the concept once more, using the house or home (*la maison*) as a heading, namely as the attraction we feel towards the intimate

7 See Fuchs 2024, pp. 127–132.
8 Fuchs 2024, p. 128.

spaces that *la maison* offers us.[9] *La maison* is the place where we experience a shelter that allows us to daydream, it is "our corner of the world", as children "our first universe", "a real cosmos in every sense of the word", and therefore "all really inhabited space bears the essence of the notion of *maison*'.[10] It is a nest, a lair, a shell. In short, the home allows us to daydream in peace, according to Bachelard, thereby integrating memories, thoughts and dreams. This is our first experience, an experience of well-being, and as such the philosopher, like the psychologist, must attend to this experience, not just to the adventures and fragmentation that take place when we are driven out into the world. The house we are born in provides us not only with a set of memories but also with habits anchored in the body. Its height – specifically in the attic and basement – creates space for our thoughts and its walls a concentration of our being. *Topo*-analysis thus becomes a *psycho*-analysis at the same time, because the house and the soul of a human being are interdependent.

Given the intimacy that characterises the house or home in Bachelard's account, it is not surprising that the contrast with the outside world is striking. Bachelard uses the experience of winter as a particular way to emphasise the difference between the home and the outside world, where the outside world essentially embodies the non-home: "And we feel warm *because* it is cold out-of-doors."[11] A "dynamic rivalry" exists between our house and the universe.[12] In this tension, the parts of the house that mediate between inside and outside will play a major role: windows, doors, thresholds, even the doorknob. "For the door is an entire cosmos of the Half-open."[13] The contrast between inside and outside is thus not really described by Bachelard as a static contrast, but as an admittedly sharp but nevertheless nuanced dialectic between here and there. What is important, however, is that the dialectical relationship between inside

9 Bachelard 2021, p. 66; Bachelard 1994, p. 12.
10 Bachelard 2021, pp. 56, 57; Bachelard 1994, pp. 4, 5.
11 Bachelard 2021, p. 96; Bachelard 1994, p. 39.
12 Bachelard 2021, p. 105; Bachelard 1994, p. 47.
13 Bachelard 2021, p. 305; Bachelard 1994, p. 222.

and outside is not symmetrical; for Bachelard, it is undoubtedly the case that the well-being that characterises *la maison* is the primary thing against which all the questionings of the outside world have to define themselves.

As several of Bachelard's critics have pointed out, his topophilia risks leading to an idyllic, not to say nostalgic, understanding of the home. Bachelard himself writes in the introduction to his book that the space of hatred and struggle hardly appears at all in its pages. Instead, it is devoted to analysing how we learn to inhabit the spaces we love. The nostalgic features of Bachelard's topophilia therefore become a critical task in the attempt to understand topophilia as a form of homesickness. Here we can turn to Michel Foucault, who in his essay on "heterotopias" has drawn attention to a tendency in phenomenology in general and in Bachelard in particular to privilege the kind of spatiality that expresses itself in demarcation and intimacy, what Foucault calls "internal space".[14] In contrast to these, he also wants to draw attention to "external space". This kind of spatiality is characterised by a threshold character, a heterogeneous space made up of places that cannot be reduced to each other and that disturbs our topophilia by forcing us out of ourselves. Topophilia, according to this critique of Bachelard, risks thinking of the home as a sphere of homogeneity, a home in which the guest, or for that matter those who live there more permanently, could never experience the traits of something alien. The home simply risks becoming, for the topophile, a sphere that never welcomes the other, but which demarcates itself against a supposedly (or sometimes truly) hostile outside world: "my home is my castle" as the saying goes.

That there are elements of such an idyllic and nostalgic understanding of home in Bachelard is quite clear, but there are also aspects of Bachelard's description of home that point in other directions. We have already seen that there is a dynamic tension between the home and the outside world, and this is concretised in all the porous openings between home and the outside world that take shape in windows and doors and which constitute genuine thresholds

14 Foucault 1986, p. 23. See Trigg 2017, pp. 5 f.

in that they mediate between inside and outside. I mentioned that the attic and the cellar create height and space, but the cellar in particular creates, at least ostensibly, a space that is beyond the domesticable parts of the house that the topophilia can master. On the one hand, the basement of the house represents the unconscious or irrational in human beings, "it is first and foremost the *dark entity* of the house, the one that partakes of subterranean forces", but on the other hand, Bachelard also claims that the unconscious "is well and happily housed".[15] If we emphasise more the first characteristic of the home, this means that the potential heterogeneity, the different and the foreign, is already present *in* the house and does not need to be invited from outside. Even if we live in a world that is in many ways enlightened, writes Bachelard, where we no longer need to carry a candle to descend into the cellar because we have electric light, the unconscious will carry a candle when it descends into the cellar; "the unconscious cannot be civilised".[16] If, on the other hand, we emphasise the second characteristic of the home, its ability to house the unconscious, a domesticated conception of the unconscious emerges, a "normal" unconscious that "knows how make itself at home everywhere".[17] Psychoanalysis comes to the rescue of the "ousted" unconscious.

How are we to understand Bachelard's description of the place of the unconscious in the house, if the unconscious is allowed to become a representative of all that can disturb the home as a place of peace and quiet? Is it in fact the case, as Dylan Trigg argues, that Bachelard's banishment of the unconscious to the basement is a way of keeping it in check?[18] If so, Bachelard's topophilia approaches a kind of metaphorical agoraphobia in its endeavour to draw a rigid line between home and away. On the other hand, it is a legitimate function of the dwelling to protect us from too raw an exposure to what threatens to invade our existence and undermine its integrity.

15 Bachelard 2021, pp. 72, 63; Bachelard 1994, pp. 18, 10.
16 Bachelard 2021, p. 74; Bachelard 1994, p. 19.
17 Bachelard 2021, p. 63; Bachelard 1994, p. 10.
18 Trigg 2017, p. 128.

The demarcation in the form of the floors of the house then becomes a way of relating to this other, without letting it master us.

The boundaries between home and away can be difficult to draw, not only between home and the outside world but also within the home itself. The endeavour to draw too rigid a line between home and away threatens to undermine even the legitimacy of topophilia as the home is emptied of space; the intimacy of too much intimacy risks becoming suffocating. As Trigg has shown in his book, aptly titled *Topophobia*, topophilia taken too far will appear suspiciously similar to agoraphobia, where the phobic suffers anxiety in public places and therefore retreats to the strictly confined and controlled home. Based on a need for control, a rigorous boundary is drawn between the home and the outside world that sharply distinguishes between the familiar and the unfamiliar. But in doing so, the home comes to be understood more in terms of a bounded shelter from the outside world than a space and a starting point for orienting oneself in the outside world. To remain a livable home, the boundary of the home with what is not home – that over there in relation to which here at home becomes just that – needs to be dynamic and porous. Just as the experience of place and space requires both poles, the relationship between home and away requires a threshold that unites by separating and vice versa.[19]

Ho omphalos

Both the height and the walls of Bachelard's *maison* can appear problematic insofar as they do not provide space for change, but threaten this space through vertical crowding and horizontal demarcation. What about another aspect of the house that is a home in Bachelard's inquiry, namely its ability to provide a point of existential orientation for human beings? He describes the home as a nest that profoundly characterises the person who inhabits it, especially if it is the childhood home. Returning to this home

19 See above, the first chapter of this book.

after years of absence means revitalising certain bodily habits and gestures, as if they were hidden in the architecture of the house.[20] Other spaces too, we might add, have a similar potential to evoke bodily memories associated with affects: the old school, the café, the neighbourhood, any space that has had an existential meaning for us through habit or significant events. Here Bachelard brings to the fore another dimension of the home that I have already highlighted, namely the qualitative distinctions that enable us to orientate ourselves in it; the Roman Forum that in antiquity it was regarded as the navel of the earth. Few images illustrate the question of existential centrality more than this one. The idea of the home as the navel of the earth provides us with a suitable starting point for understanding the meaning of the home as existential centre.

In the first chapter of *The Sacred and the Profane* from 1957, Mircea Eliade deals with the experience of sacred space in its opposition to profane space. Sacred space in its various manifestations structures space, to begin with, by introducing qualitative distinctions in relation to itself as a qualitative centre. Profane space in an absolute sense, on the other hand, would be a space without any qualitative distinctions whatsoever – it is reminiscent of the placelessness of the previous chapter – which means that differences between, for example, centre and periphery become meaningless. Profane space would therefore be the completely homogeneous space. Eliade at one point speaks of secular space as a geometrical space and writes that "no qualitative differentiation and, hence, no orientation are given by virtue of its inherent structure".[21] In contrast to secular space, sacred space provides "a fixed point and hence to acquire orientation in the chaos of homogeneity, to 'found the world' and to live in a real sense".[22] Sacred space transforms chaos into cosmos precisely by structuring the experience of spatiality. Sacred space, however, need not be identified with religious space in the strict sense, the space where religious worship is practised, such as the temple, synagogue,

20 Bachelard 2021, pp. 68 f.; Bachelard 1994, pp. 14 f.
21 Eliade 1959, p. 22.
22 Eliade 1959, p. 23.

church or the like. According to Eliade, the home and the fatherland can also be understood as sacred space in the figurative sense. What I want to emphasise here is the orientating function of these spaces in an existential sense.

In the ancient conceptions of sacred space, there is a recurring idea of a world axis or world pillar, *axis mundi*, a world tree or navel, *omphalos* or *umbilicus*, a mountain or height that unites heaven and earth and symbolises the centre of the world. "A universe comes to birth from its center; it spreads out from a central point that is, as it were, its navel."[23] By universe Eliade here means basically the same thing as a cosmos, i.e. a whole organised by means of qualitative distinctions. When a temple is built, a city constructed or a house erected, this is understood as the establishment of a universe from a centre, but in this case as a repetition of the original cosmogony, as a participation in it. Each individual dwelling, in the traditional understanding, will therefore be at the qualitative centre of the universe, its navel, because it repeats the initiation of the universe. It will have an orientating function for the individual, as well as for society. Sometimes this is expressed concretely in a pole in the middle of the house, or in a void, or more figuratively in the idea of the hearth, the open fireplace, as the centre of the house, its navel, which anchors it to the earth and to the subterranean powers, gives the house its place in the cosmos, and confers permanence and stability on it.[24] We can recall the French expression "*sans feu ni lieu*" – roughly "without hearth no home". The house, too, will thus acquire an existential orientating function; the individual houses are in turn organised around a common place: the Roman Forum. According to Eliade, all such experiences and conceptions of the house or home as a navel for the individual go back to the very experience of sacred space. Interestingly, this seems to imply that the qualitative centre of the universe can be manifested in many concrete places, and thus does not have to be identified with any geographical location.

---

23 Eliade 1959, p. 44.
24 See Vernant 2006, pp. 158 f., 174, 178–180 *et passim*.

In modern times, Eliade argues, such notions have disappeared. He sticks to a version of the history of modernisation that gives an account of how the *genius loci* has gradually come to be perceived as a metaphor and thus referred to the subjective feelings of human beings, a history that I have already recounted.[25] But he also maintains that reminiscences of the ancient approach recur in modern experiences of the fatherland, the homeland or the home. The profane space in the sense of a completely homogeneous space would be, as he says, literally impossible to orientate oneself in, since all qualitative distinctions would be meaningless: "the center renders *orientation* possible".[26] The risk, according to Eliade, is that such spaces will find it difficult to fulfil the orientating function, since "the fixed point no longer enjoys a unique ontological status" but is interchangeable.[27]

What is a fixed point for me in my private life-story is certainly different from what is a fixed point for you in your story. Sacred spaces in the modern sense are therefore not infrequently relegated to experiences of a private nature that assume a sacred status. Admittedly, even in pre-modern times, a human being undoubtedly experienced themself as a somewhat active agent in the creation of sacred spaces, but unlike modern human beings, such creation took place according to a pattern which they did not themselves originate but in which they became involved through repetition. But modern experience of the home or some other place as an orientating navel also points to the need for a place that is not residually created by an individual but in some sense also found, constituted independently of itself, given as a gift. This is because it is both unbearable and impossible to live without a point of orientation and the qualitative differences it generates. We can say that for Eliade, the whole problem I discussed in the previous chapter about *genius loci* and the relationship between place and person becomes highly relevant.

25 See here also Eliade 1991.
26 Eliade 1959, p. 63.
27 Eliade 1959, p. 23.

In concrete terms, this means that in our time it is rather the individual home that has the function of structuring space, of making it possible to orientate oneself in it. It is the navel in more than a figurative sense. In his study of the relationship between human existence and space, the phenomenologist Otto Friedrich Bollnow confirms that, even in our time, the house retains something of its sacred character. He draws mainly on Eliade, but also on Ernst Cassirer and Gerard van der Leeuw, to argue that even for us, space is not experienced as homogeneous. The house is still – with reference to Bachelard – an *imago mundi*, an image of the world.[28] If in the past it was so in a more "objective" sense, today it is more a fact we can deduce from our experience of the world. Crossing the threshold of someone's home remains an existentially meaningful act, as the home and the outside world remain qualitatively distinct from each other. There is an idea of domestic peace, of the sanctity of privacy, which is tied to the experience of home. "It is only because man lives in the house [*im Hause*]", Bollnow notes, "that he can then also be at home [*zu Hause*] in the world, and dwell in the world."[29] This sentence, which concludes Bollnow's chapter on sacred space, shows how he, too, recognises the home as a point of orientation, its function as the navel of human experience.

Orientation, being able to establish qualitative distinctions, is a central ability to avoid getting lost.[30] Lost in the forest, I look for a few points of reference that are distinct and familiar to me, which can help to dispel the impression of a homogeneous mass of trees that I find it difficult to distinguish in any meaningful way. For the more experienced forester, on the other hand, the forest appears more possible orient oneself in by virtue of the knowledge of its typical vegetation and the way it is arranged, spontaneously or cultivated. Similarly, the experience of being at home in the world means being able to orientate oneself in it. Our home is both a place and a mood – "I am home now" can mean both that I am now in the place that is

28 Bollnow 2011, pp. 137 f.
29 Bollnow 2010, p. 148; Bollnow 2011, p. 141.
30 The following is inspired by Trigg 2017, pp. 120–141.

my home and that I am overwhelmed by the affective relationship I have to home. Home – at least as the settled, affectively charged experience we have acquired in a particular place and with which we are familiar even if we no longer have access to that particular place – is essential both to my ability to relate to the outside world and to travel in it; without home in this sense, I would be placeless and have nowhere to go.

Trigg, who is certainly not a naïve topophile who elevates the home to an "idyllic fantasy", writes in his book on topophobia that "[t]he home as a centre is the principle manifestation of value and familiarity in a contingent world".[31] Again, it is about the ability to orientate oneself, not only geographically but also existentially. Feeling at home is not just a matter of locality, but a basic mood of existence that one can both have and lose and that has to do with the stability of one's own subject in relation to other people and the world. In the establishment of such a foundation, person and place are highly interrelated.

Trigg's problem in the book *Topophobia* is actually the opposite, namely when person and place are at odds with each other, as in the acute anxiety of agoraphobia, where the unknown place is perceived as threatening and triggers an alienation from one's own body. What the agoraphobic experiences, both spatially and physically, is precisely the absence of a concrete here. The fixation on the home, that this and only this really constitutes a here and not a there, follows as a concrete manifestation of the inability to orientate oneself in all that is there, outside the home. The outside world appears as without meaning and possibilities – affordances – as fundamentally inhospitable, almost unreal: "the division between home and non-home becomes a divisive distinction rather than a mutually beneficial one, with the non-home marking a threat against the stability of the home".[32] Therefore, for the agoraphobic, a sharp and stable boundary must be established between home and away.

31 Trigg 2017, p. 123.
32 Trigg 2017, p. 136.

This description of home and the world from the perspective of the agoraphobic, however, shows not only why a rigid distinction between home and the world unites Bachelard and the agoraphobic, but also the fundamental importance of the home in structuring the world in general. What counts as home for a specific individual can of course have many different answers and is not limited by Trigg to the kind of home that Bachelard writes about: *la maison*. But it becomes clear that for the person without agoraphobia, there is an expectation that the world beyond the familiar world of the home also consists of places in which it is not just chaos, in which it is possible in principle to orientate oneself. The home is the background against which we can move and feel at home in the outside world, without it falling apart into utter alienation.

Even on the thresholds between different places, where we in some sense become someone else, there is a fundamental trust in a common outside world and our own place in it. The outside world is also a place where we could live, where it would be possible to establish a new home. The fact that I have a home teaches me that others have a similar home, without the need for competition. According to Trigg, the relationship between home and the outside world, before or beyond agoraphobia, is in a fundamental sense harmonious, where the home serves to structure the outside world, making it possible to orientate oneself in it. In other words, the home is an existential navel that establishes a qualitative difference between here and there, from which we are able to orientate ourselves in the world. Paul Ricœur writes that our body "constitutes the primordial place, the here in relation to which all other places are there", but perhaps it would be more accurate to say that it is our home that establishes in our body this fundamental capacity for orientation, our existential foothold.[33]

33 Ricœur 2004, p. 43, cf. p. 149.

Die Heimat

The home I have been talking about so far is, on the whole, *la casa, la maison*. But, as I said, this is not the only use of the word home. Topophilia is not only about the affective and material ties that may exist between us and our home in this sense, and which are in some sense a prerequisite for our ability to orientate ourselves in the world. It is also about our relationship to the hometown, to the homeland, to *die Heimat*, to all the places where we can experience a sense of being at home, both in an affective sense and in our ability to orientate ourselves there. In antiquity, the individual hearths that made up the navel of private houses were thought to gather around a common navel, located in the centre of the city, which organised their interrelationships. What, then, is the meaning of this sense of being at home in a place that constitutes something more vastly expansive than *la maison*, of which this home is only a small part and in some cases may not even be located there? And to what extent can this very sense of home be a breeding ground for exclusivist and ethnocentric notions, where my *Heimat* gives rise to a topophilia that sees itself as a contrast to the topophilia of others? The concept of home – in all its meanings – refers to a phenomenon that is both a condition of and intimately associated with human existence and is therefore associated with affects that, not surprisingly, can easily give rise to conflict.

Geographer Yi-Fu Tuan's 1990 book, *Topophilia*, is largely a descriptive review of topophilia in relation to various such places and also seems in its definition to limit topophilia to precisely "the affective bond between people and place or setting".[34] Johan Asplund has therefore objected that topophilia includes not only affective relations but also volitional and material ones, and argues that Tuan underutilises the concept.[35] Asplund himself suggests that topophilia should be understood as a concept of human location in all its dimensions. There is a tendency in Tuan to portray topophilia

34 Tuan 1990, p. 4.
35 Asplund 2022, pp. 231 f.

as something we humans attribute to places through our interaction with them rather than a genuine resonant relationship where the place in question generates an approach to which topophilia is a response. For Asplund, place and person are more intimately connected than for Tuan. But at the same time, what is interesting about Tuan's book is that he makes the concept of topophilia a broad concept by allowing it to deal with a person's relationship to their material surroundings as such to the extent that it evokes or becomes the object of affective relationships.[36] Topophilia can take on many forms and is caused by many things, ranging from the visual impression of places to the possibility of orientating oneself, the sense of home and the pride of belonging to it.

In *Topophilia*, Tuan makes an important distinction between different kinds of topophilia, a distinction between what I would call concrete and abstract topophilia. Concrete topophilia can be illustrated by the farmers who have an intimate knowledge of the land they cultivate thanks to their yearly repetition of its use, for example the winemaker who harvests grapes every year to produce a wine characterised by a particular *terroir*. Everything from the geographical extent of the cultivated land to its specific quality becomes the object not only of the farmers' knowledge but also of their care. Thanks to this repetition, a farmer perceives themself as existentially bound to the soil. "The farmer's topophilia is compounded of this physical intimacy, of material dependence and the fact that the land is a repository of memory and sustains hope."[37] Topophilia in the form of patriotism, on the other hand, Tuan views with greater suspicion, because it is directed at a larger, more abstract phenomenon like the nation or the state, about which no one can have any particularly intimate knowledge. "The modern state is too large, its boundaries too arbitrary, its area too heterogeneous to command the kind of affection that arises out of experience and intimate knowledge."[38] Local patriotism is an exception here,

36 Tuan 1990, p. 93.
37 Tuan 1990, p. 97.
38 Tuan 1990, p. 100.

because it is based on “the intimate experience of place”.[39] It is, in my words, more concrete than abstract.

Of course, it is possible to know well one or more parts of a country or a city. The fact that I am familiar with Kungsportsplatsen in Gothenburg does not prevent me from feeling at home in, say, Ladugårdsgärdet in Stockholm or Möllevångstorget in Malmö. But to conclude from this that I know Sweden well would be a gross exaggeration. If a city consists of a number of neighbourhoods, and is therefore a more abstract phenomenon than a neighbourhood, then a country or a nation is even more abstract than a city or a village, because it consists – for the most part – of several cities, villages and rural areas. A nation is more an idea than a concrete phenomenon of which it is possible to have intimate knowledge. It is, to use Benedict Anderson’s phrase, more of an “imagined community” than a place we have come to know through experience.[40]

Asplund’s clarification of Tuan becomes important here: the difference between the farmer’s topophilia and the patriot’s topophilia is not only a difference in the object to which the topophilia is directed, but a difference in concreteness. If topophilia is an expression of physical location and not just affect, this means that topophilia is a function of a very concrete knowledge of place. Such knowledge is based on the multisensory experience of the infra-ordinary features of the place, which can only be acquired through habit and which is largely tacit knowledge.

It is not enough, so to speak, that I like Paris, as people from other parts of the world than Paris have done for centuries, and that I have read a lot about Paris. What I have feelings for is probably more the idea of Paris, the idea of the “romantic Paris” that has been passed on to me through literature and film. Even if I repeatedly go back to the same place in Paris, for example Place Saint-Sulpice, I would have to live there for many years to claim to know it as intimately as Perec does. Perec’s affective ties to different places in Paris also include, of course, ideas about Paris, but in his case they are combined with

39 Tuan 1990, p. 101.
40 Anderson 2016.

the intimate knowledge of those places that a lifetime in Paris can provide.[41] And even for him, "Paris" must remain something of an abstract notion, even though his knowledge of the city includes many of its places, as for us the cities we inhabit. For me, as an occasional visitor, Place Saint-Sulpice remains more a there than a here, despite my repeated visits. I simply have not acquired Perec's sense of place. I do, however, have it in other places.

To "feel at home" in a place thus means not only having an affective relationship to the place, but also an ability to orientate oneself there based on an intimate knowledge of the place acquired over time. As I pointed out earlier, a place does not become a place only by getting to know it, but it thereby becomes a place *for us*; it is the way we become part of it. Feeling at home not only in *la casa* but also in the neighbourhood around it is reflected in our language by terms like "hometown" and perhaps even more pronounced in the German *Heimat*. The latter is a controversial term, as it has often been used to express ethnocentric and exclusionary ideas about homogenous language and cultural communities. Especially in modern times, with the rise of the nation state, *Heimat* has come to be associated with an affective relationship to a country and the ethnic community with which it is associated. The *jus sanguinis* – the "right of blood" – where the right to citizenship follows the person's lineage, is a legal expression of such an idea. There are also well-known examples of how ethnicity and homeland are linked in a community of destiny, and where the homeland must therefore be defended against foreign influences that threaten to destroy it.

German novelist Andreas Maier has effectively undermined any nostalgic notions of hometown through an autofictional account in his novel *Die Heimat* of growing up in Bad Nauheim and Friedberg in Hesse in the 1970s and 1980s.[42] There was a widespread ethnocentrism, both in relation to other hometowns in the world and also within his own, which Maier depicts in a way that is both hair-raising and comical. He exposes the risk inherent in all notions of

41 See Georges Perec's posthumous work, Perec 2022.
42 Maier 2023.

Heimat, including our own. At the same time, it seems inevitable that we are all in some sense influenced by the hometowns we have inhabited. Through his autofictional narrative, Maier inscribes himself in Bad Nauheim and Friedberg; his upbringing in these very places has made him who he is.

The cultural anthropologist Ina-Maria Greverus has endeavoured to combine a critical insight into the potentially ethnocentric nature of any notion of *Heimat* with what she calls the "territorial imperative", namely that we humans are in fact constitutively bound to a place and that all humans should therefore also be given space.[43] Greverus emphasises our active ability to appropriate space, but this ability must not be equated with a desire to own, control and enclose it. Instead, it is a matter of becoming part of the space by establishing oneself in it. The place or space where someone is to feel at home must leave room for the person who is to inhabit it to make themselves at home: "*Heimat* is not 'world' but a specific environment in which people recognise themselves, are recognised and acknowledged, and feel themselves to be active, co-creating members."[44] Thus, the quest for a place-based identity and security is not only legitimate but necessary, as long as the importance of people's own creative agency in shaping their identity is also taken into account. Such agency of the person inhabiting a particular place does not at all contradict the relative primacy of the place over the person, as long as the place leaves room for the person's creative use of it together with others. The intimate knowledge of place, as for the farmer in the example above, does not mean a distanced knowledge of a fixed and ready-made, static place, but a knowledge that is involved in the shaping of place by creatively responding to its possibilities.

Our home in this extended sense is thus never finished, any more than the more private home. "No home is ever finished. We never stop building", writes Emanuele Coccia, certainly about *la casa* but

43 Greverus 1979, p. 24.
44 Greverus 1979, p. 255.

also applicable to *die Heimat*.[45] In life together, we help create our places – our homes – in all the conflicts and negotiations we are involved in, without denying – and this is my own emphasis – how places in their material, social, cultural and atmospheric form also provide us with the conditions we have to negotiate. But still, home can seem a relatively static reality, as if there could be only one home for us, whether we speak of *la maison* or *die Heimat*. What about all the nomads, all those of us who have moved, all those who are forced to migrate, or who have never had just one home but several? Are we or they no longer here at home but rather somewhere else? That is how it can be perceived, of course.

In the second chapter of this second part of the book, I gave the coordinates of my office where I wrote this chapter as latitude 57° 41' 40.6" N and longitude 11° 59' 04.6" E. But now that I am writing this chapter, that is no longer true. I have moved and find myself at latitude 59° 56' 10.2" N and longitude 10° 43' 30.0" E, or in other words in a new office for me in the Domus Theologica at the University of Oslo. Here I am surrounded by the same books as in the previous office, but in a place where I have not yet settled in, where I have not yet established myself, even though I am about to do so. Even in this globally very limited and, moreover, voluntary move, there is still a feeling of not really being here but rather there. In an atmospheric sense, moving is not the work of a moment, even when it comes to a place as existentially low-intensity as an office.[46]

According to Greverus, mobility and territoriality are not opposites, since there is always the possibility in principle of making a new home.[47] Mobility can be a way of trying to achieve a more satisfying territoriality, a home that is welcoming and inhabitable as an active act, rather than an inherent *Wanderlust* or a preference for mobile homes. If, out of necessity or because one is driven by some endeavour of one's own, one leaves for somewhere else and it is possible to set up a new home there, then the old home does not have

45 Coccia 2021, p. 72.
46 See Sigurdson 2024c, pp. 78–80.
47 Greverus 1979, p. 96.

to fall into oblivion. As cultural anthropologist Regina Römhild has pointed out in the wake of Greverus, it is not unusual for migrants to establish a kind of transnational *Heimat* where family ties and the quest for economic security are spread out between several places in different countries and that this is part of everyday life.

If migration is not perceived as a "one-way street" from one country of origin to another, it opens up the possibility of understanding it in terms of establishing networks where "people, objects, ideas and financial investments circulate".[48] Parents work in one country, children go to school in another and relatives live in a third, while spiritual, cultural and financial assets circulate across further borders – while also being part of the place where they live. This is not so much a case of "multiculturalism" as a "transnational" culture, where belonging is not defined by national territories. Making a home in the Gallus neighbourhood of Frankfurt – to use Römhild's example – becomes both a way of meeting people with similar experiences and a starting point for relationships with people in the country of origin and other countries. Transnationality, she notes, does not mean the loss of *Heimat* but the creation of a different kind of *Heimat* than the homogeneous nation state, even for those who are not migrants. Significantly, she highlights the neighbourhood as a place of participation, as a concrete form of topophilia.

Such a transnational diasporic existence is not about a late-modern cosmopolitan rootlessness, but, as Simone Weil argued, about having multiple roots, but in different places. Our human constitutive dependence on place – Greverus' territorial imperative – does not have to be cancelled just because several places can be home. The centrality of home in relation to orientation does not contradict the fact that we can make ourselves multiple centres by making ourselves at home elsewhere. For those of us who are fairly settled, travelling "over there" and then coming home can be a way of re-encountering the unfamiliar in the midst of the familiar. When we return home, it is not only the other human and animal inhabitants of the home who welcome us back, but also all the infra-

48 Römhild 2018, p. 32.

ordinary, even extraordinary things that populate it, and which we now see again; the atmosphere of domestic intimacy that we may have forgotten for a while reappears.

In general, talk of a *Heimat* risks establishing a kind of boundary more rooted in artificially created and historically evolved abstract notions than in the effect of lived experience. In *Borderlands/ La frontiera*, the poet Gloria Anzaldúa questions the idea that the mestizo experience consists in the blending of two cultures, ethnicities or nation-states, since the very idea of such a blending presupposes that these exist as established, separate, and legitimately homogeneous entities.[49] She herself believes that her home in Texas/Tejas on the US-Mexico border is a place in its own right, not a compromise between two different cultures; the plurality of her existence is in a sense original, not contingent. Her critique of certain notions of what constitutes a homeland thus does not at all imply that Anzaldúa rejects the legitimacy of the experience of home and place, of the difference between here and over there, nor that the notion of a sharp border between the United States and Mexico affects her own experience of them. What she emphatically objects to is that abstract political boundaries should be allowed to define her concrete experience of home. In principle, her experience is no more complex than any other concrete experience of home, given that heterogeneity, otherness and foreignness, is already present *in* the home and not just outside it.

In practice, of course, the politically ascribed border existence shapes Anzaldúa's experience of home as home by questioning its genuine status and legitimacy, which is the very reason for her critical questioning and objections. But this is precisely an abstract notion of home and place, of *Heimat*, versus her "intimate experience of place", as Tuan calls it, her multisensory experience of all the infra-ordinary details that characterise her home even when she is not thinking about them.[50] Although our experience of *Heimat*, of places where we feel at home, easily lends itself to abstract notions of a sharp

49 Anzaldúa 1987.
50 Tuan 1990, p. 101.

distinction between here and there and therefore also influences our experience, these notions nonetheless represent a human, all-too-human attempt to oversee and control places. Beyond this vertical displacement and horizontal demarcation, the asymmetrical remains dynamic and alien in the experience of place.

Da capo al fine

The fact that I live somewhere does not necessarily mean that I experience this dwelling as a home. And the fact that I feel at home in a place does not necessarily mean that I live there. The fact that we are talking about feeling at home in the latter case means that we are in a familiar place, a place that we have visited several times in the past, where the increased familiarity has meant that we are able to orientate ourselves there without effort, both atmospherically and geometrically.

In Gothenburg, I can orientate myself in the neighbourhoods I know well: eastern and central Gothenburg. In Oslo, the habit has not yet made me feel at home. I am reasonably familiar with the route from the Nordberg neighbourhood via Ullevål Stadion to the Blindern campus, but so far that is all. Will I also feel at home here eventually, as I do in some neighbourhoods of Gothenburg, where I still move and live? Only time will tell. Some places where I used to live for a shorter or longer period of time have lost their familiar status for me, as I no longer regularly visit them. Others will probably always feel like home, regardless of the fact that they are no longer part of my daily life. If I were to return, it is likely that bodily memories with affective significance would be awakened.

A return visit can awaken place-bound memories that in turn create a sense of belonging. But the very terminology of feeling at home sometimes appears phenomenologically limited, because there are existentially significant places – I am thinking, for example, of the workplace, but also of the café or the local pub, the sports arena or the running track, the church or the library, you can certainly add and subtract a few – which, although they cannot be called here in the sense of home, do not therefore constitute an alien there. The experience of sitting in the frequently visited and therefore perhaps also well-known and significant café is different from being at home, but at the same time essential in its own way, a third place alongside the first place of the home and the second place of the workplace, available for informal, creative interaction with others, for reflection or relaxation.[51] Similarly, the second place, the workplace, does not have to be confused with the first place, i.e. feel like home, just to avoid being impersonal. We all have many places that are meaningful to us, but in different ways. And even the existentially insignificant places, like the station or transport hub where I change buses, trams or metros, or the urban corridor I pass on my walk home, are also important in their own way, just through their impersonal mono-functionality. In many cases, it is a relief not to have to be existentially engaged by the stop where I wait for the bus.

The place becomes a place by being delimited from other places in a world that unites them. Place is therefore separated by a boundary between here and there, a boundary that is a threshold that unites by separating. But this is precisely why place can also become a genuine space in the sense that it creates the conditions for creative action. Our human attachment to place, like our physicality, is not an obstacle to our freedom but, on the contrary, its condition of possibility. By belonging to an existentially significant place or several where qualitative distinctions enable us to orientate ourselves, we can use this ability to orientate ourselves in other places, but also critically form a judgement about these distinctions between here at home and over there.

I'm probably not the only one who, after a period away from home, longs to set off on a journey, but once I'm on the move, longs for home. It is apparently a human trait to move between *Wanderlust* and *homesickness*, or as the author and cultural historian Claudio Magris says: "There is good reason why a journey is primarily a return and teaches us to dwell more unreservedly, more poetically

51 See Oldenburg 1997.

in our own home."[52] In relation to place, it is all too easy to ask whether humans are resident or nomadic when in fact we are both. Dwelling and wandering are both basic determinations of our place in the world. Dwelling is more about the familiar, about habit and permanence, wandering about the unfamiliar, the new and the changing.[53] As I noted in the second chapter of this part of the book, we need both place and space. Despite claims that place is static, I have repeatedly referred here to its dynamic nature. The difference between dwelling and wandering, I would argue, for us dwellers is not a difference between the static and the dynamic, but between the degree of viscosity: the dynamic feature of place is more viscous than wandering. For nomadic cultures, which have a tradition of being alternately sedentary and migratory, differences in the degree of viscosity necessarily take on a different form. But they too stick with their familiar, habitual and enduring places. The point of place is not to be too fluid; the vertigo of the threshold is here momentarily mitigated.

52 Magris 2018, p. 6. Cf. Caldenby 2022, p. 15, who describes how the quote from Magris became a motto for recurring city trips with architecture students in the 2010s.

53 Guzzoni 2017, p. 13.

Coda

Let us return to the twelve gods of Olympus. In some representations they are depicted as married couples, like Zeus and Hera, or as brother and sister, like Apollo and Artemis. But unlike the other five couples, Hestia and Hermes are not defined by kinship or dependency. Why then are these two arranged as a pair? According to Jean-Pierre Vernant, it is because their functions are complementary: Hestia is the god of the home or, more precisely, the god of the hearth, representing the centre of the home in Greek mythology; Hermes, on the other hand, is the god of the threshold, the god who, unlike Hestia, is always on the move and who mediates between outside and inside, living and dead, gods and men. Together, we could say, with Vernant, that they symbolically organise the ancient Greek view of spatiality:

> It could be said that, by virtue of their polarity, the Hermes-Hestia couple represents the marked tension in the archaic conception of space: space requires a center, a nodal point, with a special value, from which all directions, all qualitatively different, may be channeled and defined; yet, at the same time, space is the medium of movement, implying the possibility of transition and passage from any point to another.[1]

By now this is something we recognise. Places and thresholds; thresholds and places. Without over-interpreting the parallel, I believe that the dialectic between place and threshold returns in the existential experience of spatiality, almost as a basic human condition. We who live in a time we call modern also hold on to

1 Vernant 2006, p. 161.

powers such as Hestia and Hermes, albeit in a different way than ancient Greece, and so perhaps do most other cultures in history.

Such a movement between place and threshold is not characterised by a regular waveform in which we move back and forth between one and the other. On the contrary, it is characterised by irreversible discontinuities: crossing the threshold means, even when we eventually return to the place where we started, having become someone else. To think of existential spatiality as organised by places and thresholds is to reconcile continuity with discontinuity, not to extinguish one or the other. In our movement between places and thresholds, we limp along in an almost incalculable complexity: there are many kinds of thresholds between very different places.

The next question, therefore, is how this heterogeneity is organised and given concrete form.

Acknowledgements

In many ways, academic work is a collective effort. This book would not have been possible without friends and colleagues. I would like to thank Mikael Andersson, Michael Azar, Björn Billing, Henrik Björck, Emil Broomé, Claes Caldenby, Johan Celsing, Göran Dahlberg, Jonas Eek, Moa Ekbom, Martin Engberg, Henrik Vase Frandsen, Tonino Griffero, Tobias Hägerland, Peter Jansson, Werner Jeanrond, Maria Johansen, Per Magnus Johansson, Simone Kotva, Johan Kärnfelt, Håkan Möller, Patricia Lorenzoni, Marius Timmann Mjaaland, Cecilia Rosengren, Mats Rosengren, Monica Sand, Lydia Sandgren, Ulrich Schmiedel, Miriam Sünder, Fredrik Söderberg, Erik Törnkvist and Johan Örn. I owe my colleague Robyn Boeré a special thanks for help with language editing. Financial support has been provided by the Anna Ahrenberg Foundation, including travel grants, as well as production grants by Sven och Dagmar Saléns stiftelse. For this I am very grateful.

The first part of this book, on thresholds, has previously been published in Swedish as *Trösklar: En kritisk genomgång* (2025), and the other part, on places, as *Platser: En repetition* (2026), both by the publisher Andersson Örn. They have been edited for this translation, and I appreciate this opportunity to make them available for readers outside of Scandinavian languages.

Finally, a word of thanks to old colleagues at the Department of Literature, History of Ideas and Religion at the University of Gothenburg, and new colleagues at the Faculty of Theology, University of Oslo, which is now my permanent workplace. The conversation with old and new colleagues – but also the experience of the constant crossing of a border – has decisively influenced what I have written here. What I appreciate most, however, is the collegiality and intellectual energy, far beyond that of the academic profession, so often offered in generous measure.

Oslo in April 2026

Bibliography

Adorno, Theodor (2005). *Minima moralia: Reflections on a Damaged Life*. Translation: E. F. N. Jephcott. London/New York: Verso.

Akçay, K. Nilüfer (2019). *Porphyry's* On the Cave of the Nymphs *in its Intellectual Context*. Leiden/Boston: Brill.

Althusser, Louis (2014). "Ideology and Ideological State Apparatuses (Notes towards an Investigation)". Translation: G. M. Goshgarian. In *On the Reproduction of Capitalism: Ideology and Ideological State Apparatuses*. London/New York: Verso, pp. 232–272.

Anderson, Benedict (2016). *Imagined Communities: Reflections on the Origin and Spread of Nationalism*. Revised Edition. London: Verso.

Anzaldúa, Gloria (1987). *Borderlands/La frontiera: The New Mestiza*. San Francisco: Aunt Lute Books.

Arburg, Hans-Georg von (2011). "Doors and gates: hermeneutics and hermetics in Musil and Le Corbusier." In *Poetica*, 43:3/4, pp. 319–354.

Arendt, Hannah (1976). *The Origin of Totalitarianism*. New edition. San Diego/ New York/London: A Harvest Book.

– (1990). *On Revolution*. London: Penguin Books.

– (1998). *The Human Condition*. Second Edition. Chicago/London: University of Chicago Press.

– (2021). *Vita activa or Vom tätigen Leben*. New edition, second printing. Munich: Piper Verlag.

Aristotle (1933). *Metaphysics. Volume 1: Books 1–9*. Hugh Tredennick (ed.). The Loeb Classical Library 271. London/Cambridge, MA: Heinemann/Harvard University Press.

Asplund, Gunnar (1936). "Konst och teknik." In *Byggmästaren*. No. 14, pp. 166–171.

– (1939). "Göteborgs rådhus". In *Byggmästaren*. No. 10, pp. 157–168.

Asplund, Gunnar, Gahn, Wolter, Markelius, Sven, Paulsson, Gregor, Sundahl, Eskil, Åhrén, Uno (1980). *acceptera*. Facsimile. Arlöv: Berlings.

Asplund, Johan (2022). *Människan som samhällsvarelse: Tid, rum, individ och kollektiv*. New edition. Lund: Arkiv förlag.

Assmann, Aleida (2018). *Erinnerungsräume: Formen und Wandlungen des kulturellen Gedächtnisses*. Munich: C.H. Beck Paperback.

Assmann, Jan (2007). *Das kulturelle Gedächtnis: Schrift, Erinnerung und politische Identität in frühen Hochkulturen*. Sixth edition. Munich: C.H. Beck.

Athanasius of Alexandria (1978). *Life of St. Antony*. Translation: Robert T. Meyer. New York/Mahwah: Newman Press.

Augé, Marc (1997). *Non-Places: Introduction to an Anthropology of Supermodernity*. Translation: John Howe. London/New York: Verso.

– (2002). *In the Metro*. Translation: Tom Conley. Minneapolis: University of Minnesota Press.

Augustine (2006). *Confessions*. Translation: Henry Chadwick. Oxford World's Classics. Oxford: Oxford University Press.

Bachelard, Gaston (1994). *The Poetics of Space*. Translation: Maria Jolas. Boston: Beacon Press.

– (2021). *La poétique de l'espace*, Paris: P.U.F.

Balázs, Zoltán (2020). "Viktor Orbán's Leadership: The Prince, the Political Father, and the Doomed Trickster". In *Modern Leaders: Between Charisma and Trickery*. Agnes Horvath, Arpad Szakolczai and Manussos Marangudakis (eds.). London/New York: Routledge, pp. 191–210.

Bauman, Zygmunt (2000). *Liquid Modernity*. Cambridge/Malden: Polity Press.

Beaumont, Matthew (2016). *Nightwalking: A Nocturnal History of London*. London/New York: Verso.

Benjamin, Walter (1988). "Die Wiederkehr des Flaneurs". In *Angelus Novus: Ausgewählte Schriften 2*. Frankfurt am Main: Suhrkamp, pp. 416–421.

– (1991). *Das Passagen-Werk*. Gesammelte Schriften. Frankfurt am Main: Suhrkamp.

– (2002). *The Arcades Project*. Translation: Howard Eiland and Kevin McLaughlin. Cambridge/London: The Belknap Press.

– (2006). *Berlin Childhood around 1900*. Translation: Howard Eiland. Cambridge/London: The Belknap Press.

– (2007a). "The Work of Art in the Age of Mechanical Reproduction". In *Illuminations*. Translation: Harry Zohn. New York: Schocken Books, pp. 217–251.

– (2007b). "Theses on the Philosophy of History." In *Illuminations*. Translation: Harry Zohn. New York: Schocken Books, pp. 253–264.

Bennett, Jane (2010). *Vibrant Matter: A Political Ecology of Things*. Durham: Duke University Press.

Benta, Marius Ion (2020). "Duplicity, Corruption, and Exceptionalism in the Romanian Experience of Modernity". In *Modern Leaders: Between Charisma and Trickery*. Agnes Horvath, Arpad Szakolczai and Manussos Marangudakis (eds.). London/New York: Routledge, pp. 211–228.

Bergström, Moje (1929). "Byggnadsindustriens utveckling". In *Byggmästaren*. Häfte 1, allmänna upplagan 1, pp. 1–3.

Bergwik, Staffan (2024). *Terranautterna: Om människans dröm att upptäcka och kartlägga världen*. Stockholm: Norstedts.

Blennow, Anna (2024). *På resa med Platsens Ande*. Göteborg: Hoc Press.

Björck, Henrik (2004). *Staten, Chalmers och vetenskapen: Forskningspolitisk formering och sociala ingenjörer under Sveriges politiska industrialisering 1890–1945*. Nora: Nya Doxa.

– (2008). *Folkhemsbyggare*. Stockholm: Atlantis.

Bollnow, Otto Friedrich (2009). *Das Wesen der Stimmungen. Schriften Bd 1*. Würzburg: Königshausen & Neumann.

– (2010). *Mensch und Raum*. Eleventh Edition. Stuttgart: Kohlhammer.

– (2011). *Human Space*. Translation: Christine Shuttleworth. London: Hyphen Press.

Bourdieu, Pierre (2002). *Outline of a Theory of Practice*. Translation: Richard Price. Cambridge: Cambridge University Press.

Braidwood, Ella (2023). "Life on the edge: meet the man who walked around the UK". In *The Guardian*, Sunday 5 February, https://www.theguardian.com/lifeandstyle/2023/feb/05/life-on-the-edge-meet-the-man-who-walked-around-the-uk (retrieved 240918).

Buck-Morss, Susan (1989). *The Dialectics of Seeing: Walter Benjamin and the Arcades Project*, Cambridge: MIT Press.

Buzard, James (2001). "Perpetual Revolution". In *Modernism/modernity*. 8:4, pp. 559–581.

Böhme, Gernot (2013). *Architektur und Atmosphäre*. Second edition. Munich: Wilhelm Fink Verlag.

Caldenby, Claes, (ed.) (2010). *Asplunds rådhus i Göteborg*. Stockholm: Museum of Architecture.

– (2022). *Städerna och minnet*. Göteborg: Arche Press.

– (2023). "En arkitekt och hans tid". In *Arvid Fuhre: Arkitekt och kulturbärare*. Claes Caldenby with Ingrid Sommar, Krister Engström and Louise Hedenberg. Stockholm: Appell förlag, pp. 74–204.

Carter, Jason (2010). "The notion of Threshold: An investigation into conceptual accompaniment in Aristotle and Hegel". In *Conserveries mémorielles: Revue transdisciplinaire*, no 7: Seuils, Thresholds, Soglitudes. http://journals.openedition.org/cm/431 (retrieved 240918).

Casey, Edward S. (1998). *The Fate of Place: A Philosophical History*. Berkeley/Los Angeles/London: University of California Press.

Certeau, Michel de (1988). *The Practice of Everyday Life*. Translation: Steven Rendall. Berkeley/Los Angeles/London: University of California Press.

Childs, Marquis (1936). *Sweden: The Middle Way*. New Haven: Yale University Press.

Churchill, Winston (1943). "House of Commons Rebuilding". Volume 393: Debated on Thursday 28 October 1943. https://hansard.parliament.uk/commons/1943-10-28/debates/4388c736-7e25-4a7e-92d8-eccb751c4f56/HouseOfCommonsRebuilding (accessed 250314).

Cicero, Marcus Tullius (1967). *De oratore I, II*. Loeb Classical Library. Translation: E. W. Sutton. Cambridge: Harvard University Press.

Coccia, Emanuele (2021). *Filosofia della casa: Lo spazio domestico e la felicità*. Turin: Einaudi.

Connerton, Paul (1989). *How Societies Remember*. Cambridge: Cambridge University Press.

– (2009). *How Modernity Forgets*. Cambridge: Cambridge University Press.

Cosnard, Denis (2022). *Paris de Georges Perec: La ville mode d'emploi*. Paris: Parigramme.

Crary, Jonathan (2013). *24/7: Late Capitalism and the Ends of Sleep*. London: Verso.

Dahlberg, Helena (2013). *Vad är kött? Kroppen och människan i Merleau-Pontys filosofi*. Göteborg: Glänta Produktion.

Douglas, Mary (2001). *Purity and Danger: An Analysis of the Concepts of Pollution and Taboo*. London/New York: Routledge.

Dovey, Kim (1999). *Framing Places: Mediating Power in Built Form*. London/New York: Routledge.

Eisenstadt, S. N. (1978). *Revolution and the Transformation of Societies: A Comparative Study of Civilisations*. New York: The Free Press.

– (1999). *Fundamentalism, Sectarianism, and Revolution: The Jacobin Dimension of Modernity*. Cambridge: Cambridge University Press.

Eliade, Mircea (1965). *The Myth of the Eternal Return: Or, Cosmos and History*. Translation: Willard R. Trask. Princeton: Princeton University Press.

– (1959). *The Sacred and the Profane: The Nature of Religion*. Translation: Willard R. Trask. New York: Harcourt Brace Jovanovich.

Elliot, Hans O. (1929). "Moderna rationaliseringstendenser inom husbyggnadskonsten, icke sällan på gränsen till det irrationella och stundom överskridande densamma". In *Byggmästaren*. No. 15, general edition 10, pp. 73–76.

Ellul, Jacques (1964). *The Technological Society*. Translation: John Wilkinson. New York: Vintage Books.

"Erfahrung" (1972). In *Historisches Wörterbuch der Philosophie*. Volume 2. Joachim Ritter and Karlfried Gründer (eds.). Basel: Schwabe, col. 609–617.

Eriksson, Eva (2001). *Den moderna staden tar form: Arkitektur och debatt 1910–1935*. Stockholm: Ordfront, 2001.

– (2010). "Gunnar Asplund." In *Asplunds rådhus i Göteborg*. Claes Caldenby (ed.). Stockholm: Museum of Architecture, pp. 31–67.

Ernaux, Annie (2017). *The Years*. Translation: Alison L. Strayer. New York: Seven Stories Press.

Foucault, Michel (1986). "Of Other Spaces". In *Diacritics*, 16:1, pp. 22–27.

– (1994). "Des espaces autres". In *Dits et écrits: 1954–1988*. Vol. 4: 1980–1988. Paris: Gallimard, pp. 752–762.

Frendin, Torsten (1936). "Funktionalismen i verkligheten" (review). In *Byggmästaren*. No. 10, pp. 111–118.

Freud, Sigmund (1982). "Erinnern, Wiederholen und Durcharbeiten (Weitere Ratschläge zur Technik der Psychoanalyse II)". In *Schriften zur Behandlungstechnik*. Frankfurt am Main: Fischer Taschenbuch Verlag, pp. 207–215.

– (1981a). *An Outline of Psycho-Analysis*. In *The Standard Edition*. Vol. XXIII (1937–1939). Translation: James Strachey. London: The Hogarth Press, pp. 144–207.

– (1981b). "Analysis Terminable and Interminable". In *The Standard Edition*. Vol. XXIII (1937–1939). Translation: James Strachey. London: The Hogarth Press, pp. 216–253.

– (1981c). "Remembering, Repeating and Working-Through (Further Recommendations on the Technique of Psycho-Analysis: I)". In *The Standard Edition*. Vol. XII (1911–1913). Translation: James Strachey. London: The Hogarth Press, pp. 147–156.

Friberger, Erik, and Blomquist, Ture (1932). *Bostadssocialism: En uppgift för socialistisk ungdom*. Göteborg: Odinsala studiekoloni.

Frykholm, Hannes (2025). *Infra-Ordinary Architecture*. Stockholm/Umeå: Andersson Örn.

Fuchs, Thomas (2024). *Verkörperte Gefühle: Zur Phänomenologie von Affektivität und Interaffektivität*. Frankfurt am Main: Suhrkamp Verlag.

Gadamer, Hans-Georg (2006). *Truth and Method*. Second, revised edition. Translation: Joel Weinsheimer and Donald G. Marshall. London/New York: Continuum.

Gassenhuber, Rudolf, and Kozljanič, Robert Josef (2024). *In Search of Lost Liveliness: Nature, Dominance, Sense of Heart – The Long Way to Friendship with Nature, A Dialogue in Letters*. Translation into English: Richard Falzmann. Milan: Mimesis International.

Gedin, Per I. (2018). *När Sverige blev modernt: Gregor Paulsson, Vackrare vardagsvara, funktionalismen och Stockholmsutställningen 1930*. Stockholm: Albert Bonniers Förlag.

Gennep, Arnold van (1960). *Rites of Passage*. Translation: Monika B. Vizedom and Gabrielle L. Caffee. Chicago: University of Chicago Press.

– (1981). *Les rites de passage*. Paris: Éditions A. et J. Picard.

Gertz, Nolen (2024). *Nihilism and Technology*. Updated edition. Lanham/Boulder/New York/London: Rowman & Littlefield.

Gibson, James J. (1986). *The Ecological Approach to Visual Perception*. New York/Hove: Psychology Press.

Goffman, Erling (1956). *The Presentation of Self in Everyday Life*. Edinburgh: University of Edinburgh Social Sciences Research Centre.

Gramsci, Antonio (1992). *Selections from the Prison Notebooks*. Translation: Quintin Hoare and Geoffrey Nowell Smith. New York: International Publishers.

Greverus, Ina-Maria (1979). *Auf der Suche nach Heimat*. München: C.H. Beck.

Griffero, Tonino (2017). *Quasi-Things: The Paradigm of Atmospheres*. Translation: Sarah De Sanctis. New York: SUNY Press.

– (2020). *Places, Affordances, Atmospheres: A Pathic Aesthetic*. Oxon/New York: Routledge.

– (2021). *The Atmospherical 'We': Moods and Collective Feelings*. Milan: Mimesis International.

Gropius, Walter (1931). "Byggnadsväsendets rationalisering". In *Byggmästaren*. Issue 28, general edition 19, pp. 149–156.

Gros, Frédéric (2009). *Marcher, une philosophie*. Paris: Carnets Nord.

Guzzoni, Ute (2017). *Wohnen und Wandern.* Munich: Karl Alber Publishing House.

Halbwachs, Maurice (1968). *La mémoire collective.* Second edition. Paris: Presses universitaires de France.

– (1980). *The Collective Memory.* Translation: Francis J. Ditter and Vida Yazdi Ditter. New York: Harper & Row.

– (1992). *On Collective Memory.* Translation: Lewis A. Coser. Chicago/London: University of Chicago Press.

Haraway, Donna J. (2016). *Staying with the Trouble: Making Kin in the Chthulucene.* Durham: Duke University Press.

Harries, Karsten (1997). *The Ethical Function of Architecture*, Cambridge/London: The MIT Press.

– (2001). *Infinity and Perspective*, Cambridge/London: The MIT Press.

Harvey, David (1990). *The Condition of Postmodernity: An Enquiry into the Origins of Cultural Change.* Cambridge/Oxford: Blackwell.

Hasse, Jürgen (2015). *Der Leib der Stadt: Phänomenographische Annäherungen.* Freiburg/Munich: Verlag Karl Alber.

– (2017). *Die Aura des Einfachen: Mikrologien räumlichen Erlebens.* Band 1. Munich: Verlag Karl Alber.

– (2020). *Wohnungswechsel: Phänomenologie des Ein- und Auswohnens.* Bielefeld: transcript Verlag.

Hayden, Dolores (1977). "Skyscraper Seduction – Skyscraper Rape". In *Heresies*, 1:2, pp. 108–115.

Hazan, Eric (2011). *The Invention of Paris: A History in Footsteps.* Translation: David Fernbach. London/New York: Verso.

– (2016). *Une traversée de Paris.* Paris: Éditions du Seuil.

– (2021). *Le tumulte de Paris.* Paris: Éditions La Fabrique.

Hegel, G. W. F. (1979). *Enzyklopädie der philosophischen Wissenschaften im Grundrisse 1830*, vol. 3. Werke 10. Frankfurt am Main: Suhrkamp.

Heidegger, Martin (1975). "Building Dwelling Thinking". In *Poetry, Language, Thought*. Translation: Albert Hofstadter. New York: Harper & Row, pp. 145–161.

– (2000). "Bauen Wohnen Denken". *Vorträge und Aufsätze*. Gesamtausgabe. Frankfurt am Main: Vittorio Klostermann, pp. 147–164.

Horvath, Agnes (2013). *Modernism and Charisma*. Basingstoke/New York: Palgrave Macmillan.

Horvath, Agnes, and Szakolczai, Arpad (2018). *Walking into the Void: A Historical Sociology and Political Anthropology of Walking*. London/New York: Routledge.

– (2020). *The Political and Sociological Anthropology of Evil: Tricksterology*. London/New York: Routledge.

Horvath, Agnes, Szakolczai, Arpad, and Marangudakis, Manussos, (eds. 2020). *Modern Leaders: Between Charisma and Trickery*. London/New York: Routledge.

Homer (2016). *The Odyssey*. Translation: Anthony Verity. Oxford World's Classics. Oxford: Oxford University Press.

Ingold, Tim (2016). *Lines: A Brief History*. New edition. London/New York: Routledge.

Jameson, Fredric (1991). *Postmodernism, or, The Cultural Logic of Late Capitalism*. Durham: Duke University Press.

Jansson, Peter (2022). "Har du handlat i enlighet med ditt begär? Några reflektioner över psykoanalysens mål". In *Divan*. No. 1–2, pp. 136–147.

Jaspers, Karl (2019). *Psychologie der Weltanschauungen*. Karl Jaspers Gesamtausgabe. Volume I/6. Oliver Immel (ed.). Basel: Schwabe.

Joas, Hans (2017). *Die Macht des Heiligen: Eine Alternative zur Geschichte von der Entzauberung*. Frankfurt am Main: Suhrkamp.

Johansson, Gotthard (1930). "Konstindustrien på Stockholmsutställningen". In *Ord & Bild*. Vol. 39, pp. 485–496.

– (1931). *Funktionalismen i verkligheten*. Stockholm: Albert Bonniers Förlag.

– (1933). "Chikagoutställningen och den moderna arkitekturen i Amerika". In *Byggmästaren*. Booklet 31, general edition 21, pp. 153–159.

– (1935) *Funktionalismens framtid: Kritiska betraktelser*. Stockholm: Wahlström & Widstrand.

Jonsson, Karin (2022). "Revolution". In *Svenska begreppshistorier: Från antropocen till åsiktskorridor*. Jonas Hansson and Kristiina Savin (eds.). Stockholm: Fri Tanke, pp. 581–591.

Jonsson, Stefan (2000). *Subject Without Nation: Robert Musil and the History of Modern Identity*. Durham/London: Duke University Press.

Jordheim, Helge (2023). *Berlin: Byen og historien*. Oslo: Forlaget Press.

Kant, Immanuel (1979). *The Conflict of the Faculties*. Translation: Mary J. Gregor. New York: Abaris Books.

Kierkegaard, Søren (1964). *Repetition: An Essay in Experimental Psychology*. Translation: Walter Lowrie. New York: Harper & Row.

Kirkland, Stephane (2013). *Paris Reborn: Napoléon III, Baron Haussmann, and the Quest to Build a Modern City*. New York: St Martin's Press.

Koolhaas, Rem, Trüby, Stephan, Westcott, James, and Petermann, Stephan, (eds. 2018). *Elements of Architecture*. Cologne: Taschen.

Koselleck, Reinhart (1989). "Historische Kriterien des neuzeitlichen Revolutionsbegriff". In *Vergangene Zukunft: Zur Semantik geschichtlicher Zeiten*. Frankfurt am Main: Suhrkamp, pp. 67–86.

Kozljanič, Robert Josef (2004a). *Der Geist eines Ortes: Kulturgeschichte und Phänomenologie des Genius Loci. Bd 1.: Antike–Mittelalter*. München: Albunea Verlag.

– (2004b). *Der Geist eines Ortes: Kulturgeschichte und Phänomenologie des Genius Loci. Bd 2.: Neuzeit–Gegenwart*. München: Albunea Verlag.

Kärnekull, Olof (1931). "Om byggnadsindustrien och rationaliseringen". Issue 13, general edition 9, pp. 65–72.

Lacan, Jacques (1998). *The Four Fundamental Concepts of Psychoanalysis*. The Seminar of Jacques Lacan. Book XI. Translation: Alan Sheridan. New York/London: W. W. Norton.

Landzelius, Michael (2009). "Spatial reification, or, collectively embodied amnesia, aphasia, and apraxia". In *Semiotica*, 175:1, pp. 39–75.

Latour, Bruno (1992) "Where Are the Missing Masses? The Sociology of a Few Mundane Artifacts". In *Shaping Technology/Building Society: Studies in Sociotechnical Change*. W. E. Bijker and John Law (eds.). Cambridge: The MIT Press, pp. 225–258.

Lauretis, Teresa de (2008). *Freud's Drive: Psychoanalysis, Literature and Film.* Basingstoke: Palgrave Macmillan.

Lear, Jonathan (2010). *Freud.* New York/London: Routledge.

– (2008). *Radical Hope: Ethics in the Face of Cultural Devastation.* Cambridge/London: Harvard University Press.

Lee, Vernon (2021). *Genius Loci: Notes on Places.* Milton Keynes: Adventures in Ideas.

Lefebvre, Henri (1991). *The Production of Space.* Translated by Donald Nicholson-Smith. Oxford/Cambridge: Blackwell.

Lewis, Christian (2023). *Finding Hildasay: How one man walked the UK's coastline and found hope and happiness.* London: Macmillan.

Liedman, Sven-Eric (2015). *Karl Marx: En biografi.* Stockholm: Albert Bonniers Förlag.

Lo-Johansson, Ivar (1976). *Författaren: Självbiografisk berättelse.* Stockholm: Albert Bonniers förlag.

– (1982). *Tröskeln.* Stockholm: Albert Bonniers förlag.

Lynch, Kevin (1960). *The Image of the City.* Cambridge/London: The MIT Press.

Magris, Claudio (2018). *Journeying.* Translation: Anne Milano Appel. New Haven/London: Yale University Press.

Maier, Andreas (2023). *Die Heimat: Novel.* Berlin: Suhrkamp Verlag.

"La Maison d'Église de la Défense" (2025). https://ndp92.fr (retrieved 250314).

Malpas, Jeff (2006). *Heidegger's Topology: Being, Place, World.* Cambridge: MIT Press.

– (2012). *Heidegger and the Thinking of Place: Explorations in the Topology of Being.* Cambridge/London: The MIT Press.

– (2018). *Place and Experience: A Philosophical Topography.* Second edition. London/New York: Routledge.

Marcus, Lars (2023). *Städernas stenar: Hur den byggda staden formar den levda staden.* Stockholm: Dokument press.

Marion, Jean-Luc (2002a). *Being Given: Toward a Phenomenology of Givenness*. Translation: Jeffrey L. Kosky. Stanford: Stanford University Press.

– (2002b). *In Excess: Studies of Saturated Phenomena*. Translation: Robyn Horner and Vincent Berraud. New York: Fordham University Press.

– (2008). *The Visible and the Revealed*. Translation: Jeffrey L. Kosky. New York. Fordham University Press.

Marrinan, Michael (2009). *Romantic Paris: Histories of a Cultural Landscape, 1800-1850*. Stanford: Stanford University Press.

Marx, Karl, and Engels, Friedrich (2002). *The Communist Manifest*. Translation: Samuel Moore. London: Penguin Books.

Marx, Karl. (2000). "The Eighteenth Brumaire of Louis Bonaparte". In *Karl Marx: Selected Writings*. David McLellan (ed.). Oxford: Oxford. University Press, pp. 329–355.

Menninghaus, Winfried (1986). *Schwellenkunde: Walter Benjamin's Passage des Mythos*. Frankfurt am Main: Suhrkamp.

Merleau-Ponty, Maurice (2005). *The Phenomenology of Perception*. Translation: Colin Smith. London/New York: Routledge.

Mumford, Lewis (1961). *The City in History: Its Origins, Its Transformations, and Its Prospects*. San Diego/New York/London: Harvest.

Murdoch, Iris (2014). *The Sovereignty of Good*. London/New York: Routledge.

Musil, Robert (2006) "Doors and Portals". In *Posthumous Papers of a Living Author*. Translation: Peter Wortsman. Brooklyn: Archipelago Press, pp. 59–63.

Myrdal, Alva and Gunnar (1997). *Kris i befolkningsfrågan*. Nora: Nya Doxa.

Myrdal, Gunnar (1932). "Socialpolitikens dilemma", I–II. In *Spektrum*. 3–4, pp. 1–13, 13–31.

Nietzsche, Friedrich (1998). *On the Genealogy of Morals*. Oxford World's Classics. Translation: Douglas Smith. Oxford: Oxford University Press.

Nora, Pierre (1989). "Between Memory and History: *Les Lieux de Mémoires*". In *Representations*, 26, pp. 7–24.

Norberg-Schulz, Christian (1980). *Genius Loci: Towards a Phenomenology of Architecture*. New York: Rizzoli.

Nordström, Ludvig (1938). *Lort-Sverige*. Third edition. Stockholm: Kooperativa förbundets bokförlag.

Näsström, Gustaf (1930). *Svensk funktionalism*. Stockholm: Natur & Kultur.

Oldenburg, Ray (1997). *The Great Good Place: Cafés, Coffee Shops, Bookstores, Bars, Hair Salons, and Other Hangouts at the Heart of a Community*. Philadelphia: Da Capo Press.

Otto, Rudolf (2014). *Das Heilige: Über das Irrationale in der Idee des Göttlichen und sein Verhältnis zum Rationalen*. München: C. H. Beck.

Ovidius Naso, Publius (2011). *Fasti*. Translation: Anne and Peter Wiseman. Oxford World's Classics. Oxford: Oxford University Press.

Pallasmaa, Juhani (2017). "Place and Atmosphere". In *The Intelligence of Place: Topographies and Poetics*. Jeff Malpas (ed.). London/New York: Bloomsbury, pp. 129–155.

"Paris La Défense, bien plus qu'un quartier d'affaires" (2025). https://parisladefense.com/en/discover/towers/paris-la-defense-arena (retrieved 250314).

Paulsson, Gregor (1941). "Byggnaden och människan". In *Byggmästaren*. No. 13, pp. 31–42.

– (1945). "Under 100 år". In *Form: Svenska slöjdföreningens tidskrift*. 41:1, pp. 3–7.

Perec, Georges (1997). *Species of Spaces and Other Pieces*. Translation: John Sturrock. London: Penguin.

– (2010). *An Attempt at Exhausting a Place in Paris*. Translation: Marc Lowenthal. Cambridge: Wakefield Press.

– (2022). *Lieux: Inédit*. Paris: Éditions du Seuil.

Perullo, Nicola (2021). *Epistenology: Wine as Experience*. New York: Columbia University Press.

Pesch, Rudolf (1986). *Die Apostelgeschichte*. 1. Teilband: Apg 1–12. Evangelisch-Katholischer Kommentar zum Neuen Testament. V/1. Zurich/Einsiedeln/Cologne/Neukirchen-Vluyn: Benziger Verlag/Neukirchener Verlag.

Pickstock, Catherine (2013). *Repetition and Identity*. Oxford: Oxford University Press.

Polanyi, Michael (2009). *The Tacit Dimension*. Chicago/London: University of Chicago Press.

Ponzi, Mauro (2014). "Neapel als Topographie der Zwischenräume: Die Schwelle zwischen Altem und Neuem bei Walter Benjamin". In *Schwellen: Ansätze für eine neue Theorie des Raums*. Sieglinde Borvitz and Mauro Ponzi (eds.). Düsseldorf: Düsseldorf University Press, pp. 101–120.

Porphyry (1986). *L'antro delle Ninfe*. Laura Simonini (ed.). Milan: Adelphi Edizioni.

Quivigier, Pierre-Yves (2023). *Une philosophie du vin: Millésime 2023*. Paris: Albin Michel.

Radin, Paul (1972). *The Trickster: A Study in American Indian Mythology*. New York: Schocken Books.

Ravaisson, Félix (2008). *On Habit*. Translation: Clare Carlisle and Mark Sinclair. London/New York: Continuum.

Reitzel, Erik (1984). "Le cube ouvert: Structures and Foundations". In *Tall Buildings: Proceedings of The International Conference on Tall Buildings, 22–26 October 1984, Singapore*. S. D. Ramaswamy and C. T. Tam (eds.). Singapore: IES, pp. 191–197.

"Revolution, Rebellion, Aufrühr, Bürgerkrieg" (2004). In *Geschichtliche Grundbegriffe: Historisches Lexikon zur politisch-sozialen Sprache in Deutschland*, Vol. 5: Pro-Soz. Otto Brunner, Werner Conze and Reinhart Koselleck (eds.). Stuttgard: Klett-Cotta, pp. 653–789.

Ricœur, Paul (1976). *Interpretation Theory: Discourse and the Surplus of Meaning*. Fort Worth: The Texas Christian University Press.

– (2004). *Memory, History, Forgetting*. Translation: Kathleen Blamey and David Pellauer. Chicago: University of Chicago Press

Robinson, Jancis, and Harding, Julia, (eds. 2015). *The Oxford Companion to Wine*. Fourth edition. Oxford: Oxford University Press.

Rosa, Hartmut (2020). *Resonanz: Eine Soziologie der Weltbeziehung*. Third edition. Frankfurt am Main: Suhrkamp Verlag.

Rose, Gillian (1995). *Love's Work*. London: Chatto & Windus.

Rossi, Aldo (1984). *The Architecture of the City*. Translation: Diane Gherardi and Joan Ockman. Cambridge/London: The MIT Press.

Ruin, Hans (2018). *Being With the Dead: Burial, Ancestral Politics, and the Roots of Historical Consciousness*. Stanford: Stanford University Press.

Ruskin, John (2009). "The Nature of Gothic". In *Selected Writings*. Ed. Dinah Birch. Oxford World's Classics. Oxford: Oxford University Press, s. 32–63.

Rykwert, Joseph (1976). *The Idea of a Town: The Anthropology of Urban Form in Rome, Italy and the Ancient World*. London: Faber and Faber.

Römhild, Regina (2018). "Global *Heimat*: (Post)Migrant Productions of Transnational Space". In *Anthropological Journal of European Cultures*, vol. 27:1, pp. 27–39.

Sakwa, Richard (2015). "Liminality and Postcommunism." In *Breaking Boundaries: Varieties of Liminality*, Agnes Horvath, Bjørn Thomassen, and Harald Wydra (eds.). New York/Oxford: Berghahn, pp. 205–225.

Sand, Monica (2008). *Konsten att gunga: Experiment som aktiverar mellanrum*. Stockholm: Axl Books.

Schmitz, Hermann (2019a). *Der Gefühlsraum. System der Philosophie, vol. III/2*. Munich: Verlag Karl Alber.

– (2019b). *Das Göttliche und der Raum. System der Philosophie, vol. III/4*. München: Verlag Karl Alber.

Sennett, Richard (1994). *Flesh and Stone: The Body and City in Western Culture*. London/Boston: Faber and Faber.

– (1998). *The Corrosion of Character: The Personal Consequences of Work in the New Capitalism*. New York: W. W. Norton.

Siegert, Bernhard (2012). "Doors: On the Materiality of the Symbolic". Translation: John Durham Peters. In *Grey Room*. 47, pp. 6–23.

– (2024). *Final Frontiers: Eine Medienarchäologie des Meeres*. Paderborn: Brill Fink.

Sigurdson, Ola (2000). *Den lyckliga filosofin: Etik och politik hos Hägerström, Tingsten, makarna Myrdal och Hedenius*. Stockholm/Stehag: Brutus Östlings bokförlag Symposion.

– (2012). *Theology and Marxism in Eagleton and Žižek: A Conspiracy of Hope*. New York: Palgrave Macmillan.

– (2013). "Žižek, the Death Drive, and Zombies: A Theological Account". In *Modern Theology*, 29:3, pp. 361–380.

– *Heavenly Bodies: Incarnation, the Gaze, and Embodiment in Christian Theology*. Translation: Carl Olsen. Grand Rapids: Eerdmans.

– (2019). "Only Vulnerable Creatures Suffer: On Suffering, Embodiment and Existential Health". In *Phenomenology of the Broken Body*. Espen Dahl, Cassandra Falke and Thor Eirik Eriksen (eds.). London/New York: Routledge, pp. 87–100.

– (2021). *Gudomliga komedier: Humor, subjektivitet, transcendens. Vol. 3: Kritik av den existentiella humorn*. Göteborg: Glänta.

– (2024a). "Flickläroverket: En passage". In *En båge genom tiden – ritualer kring en Göteborgshistoria: Om Flickläroverket i Artisten*. Monica Sand (ed.). Göteborg: Korpen, pp. 23–30.

– (2024b). "Göteborg: En jämförelse". In *Tidens tecken*, pp. 85–96.

– (2024c). *Atmospheres in Everyday Life: On Existential Spatiality*. "Atmospheric Spaces". Milan: Mimesis International.

– (2025a). "Den liminala historien: Avbruten passagerit i Donna Tartts mysterieroman". In *Oavsett vad vi kallar skönhet: Studier av Donna Tartts roman Den hemliga historien*. Moa Ekbom and Ingela Nilsson (eds.). Stockholm: Appell förlag, pp. 93–116.

– (2025b). "'Sankt Kristoffer, vaka över oss': Om tillit under liminalitetens tecken". In *Arche: Tidskrift för psykoanalys, humaniora, konst och arkitektur*, nr 94–95, pp. 183–195.

– (2026). "Is Home a Starting Point or an End Point? Towards a Hermeneutics of Dwelling". In *Engaging Werner G. Jeanrond's Theological Thinking: Starting Points*. Ulrich Schmiedel, Ola Sigurdson, and Jayne Svenungsson (eds.). London: Bloomsbury T&T Clark, pp. 61–74.

Simmel, Georg (1994). "Bridge and Door". Translation: Mark Ritter. In *Theory, Culture & Society*. 11:1, pp. 5–10.

– (2020). *Die Großstädte und das Geistesleben*. Second edition. Frankfurt am Main: Suhrkamp.

Simson, Otto von (1989). *The Gothic Cathedral: Origins of Gothic Architecture and the Medieval Concept of Order*. Expanded edition. Princeton: Princeton University Press.

Søndergaard, Morten (2020). *Forsøg på at aflytte et sted i Paris*. Århus: Herman & Frudit.

Solnit, Rebecca (2000). *Wanderlust: Walking on foot*. New York: Viking Books.

Spence, Jonathan D. (1985). *The Memory Palace of Matteo Ricci*. London: Faber.

Stoker, Bram (2011). *Dracula*. Revised Edition. London: Penguin Classics.

Ströker, Elisabeth (1977). *Philosophische Untersuchungen zum Raum*. Second edition. Frankfurt am Main: Klostermann.

Szakolczai, Arpad (2000). *Reflexive Historical Sociology*. London/New York: Routledge.

– (2015). "Liminality and Experience: Structuring Transitory Situations and Transformative Events". In *Breaking Boundaries: Varieties of Liminality*, Agnes Horvath, Bjørn Thomassen and Harald Wydra (eds.). New York/Oxford: Berghahn, pp. 11–38.

– (2017). *Permanent Liminality and Modernity: Analysing the Sacrificial Carnival through Novels*. London/New York: Routledge.

Tally, Robert T. (2013). *Spatiality*. Oxon/New York: Routledge.

Thibaud, Jean-Paul (2015). *En quête d'ambiances: Éprouver la ville en passant*. Genève: MètisPresses.

Thomassen, Bjørn (2014). *Liminality and the Modern: Living Through the In-Between*. Farnham/Burlington: Ashgate.

Trigg, Dylan (2012). *The Memory of Place: A Phenomenology of the Uncanny*. Athens: Ohio University Press.

– (2017). *Topophobia: A Phenomenology of Anxiety*. London: Bloomsbury.

Trumbull, H. Clay (1896). *The Threshold Covenant, or, The Beginning of Religious Rites*. New York: Charles Scribner's Sons.

Tuan, Yi-Fu (1977). *Space and Place: The Perspective of Experience*. Minneapolis/London: University of Minnesota Press.

– (1990). *Topophilia: A Study of Environmental Perception, Attitudes, and Values*. New York: Columbia University Press.

Turner, Victor (1967). "Betwixt and Between: The Liminal Period in *Rites de Passage*". In *The Forest of Symbols: Aspects of Ndembu Ritual*. Ithaca: Cornell University Press, pp. 93–111.

– (1974). "Liminal to Liminoid, in Play, Flow, and Ritual: An Essay in Comparative Symbology". In *Rice Institute Pamphlet – Rice University Studies*, 60. No. 3, pp. 53–92.

– (2017). *The Ritual Process: Structure and Anti-Structure*. London/New York: Routledge.

Turner, Victor, and Turner, Edith (1978). *Image and Pilgrimage in Christian Culture: Anthropological Perspectives*. New York: Columbia University Press.

Velicu, Adrian (2010). *Civic Cathecisms and Reason in the French Revolution*. Farnham/Burlington: Ashgate.

Vernant, Jean-Pierre (2006). *Myth and Thought among the Greeks*. Translation: Janet Lloyd with Jeff Fort. New York: Zone Books.

Virgil (2007). *Aeneid*. Translation: Frederick Ahl. Oxford: Oxford University Press.

Waldenfels, Bernhard (2013). *Sinnesschwellen: Studien zur Phänomenologie des Fremden 3*. Third edition. Frankfurt am Main: Suhrkamp.

– (2018). *Grundmotive einer Phänomenologie des Fremden*. Sixth edition. Frankfurt am Main: Suhrkamp.

Walker, Enrique (2001). "Paul Virilio on Georges Perec" (interview). In *AA Files*, no 45/46, pp 15–18.

Weber, Max (1978). *Economy and Society: An Outline of Interpretative Sociology*. Berkeley/Los Angeles/London: University of California Press.

Weil, Simone (1952). *The Need for Roots*. Translation: A. F. Vills. London: Routledge & Kegan Paul.

Wiklund, Martin (2006). *I det modernas landskap: Historisk orientering och kritiska berättelser om det moderna Sverige mellan 1960 och 1990*. Stockholm/Stehag: Brutus Östlings bokförlag Symposion.

Wittgenstein, Ludwig (1986). *Philosophical Investigations*. Third edition. Translation: G. E. M. Anscombe. Oxford: Basil Blackwell.

Yates, Frances A. (2014). *The Art of Memory*. London: The Bodley Head.

Zetterholm, Magnus (2009). *Approaches to Paul: A Student's Guide to Recent Scholarship*. Minneapolis: Fortress Press.

Žižek, Slavoj (2008). *In Defence of Lost Causes*. London/New York: Verso.

Åhrén, Uno (1925). "Brytningar." In *Svenska slöjdföreningens årsbok*. Vol. 21, pp. 7–36.

– (1928). "Synpunkter på stadsbyggandet". In *Byggmästaren*. Issue 32, general edition 22, pp. 173–175.

– (1930). "Bostäderna på Stockholmsutställningen". In *Ord & Bild*. Vol. 39, pp. 429–437.

– (1936). "Konst och teknik". In *Byggmästaren*. No. 14, pp. 172–176.

MIMESIS GROUP
www.mimesis-group.com

MIMESIS INTERNATIONAL
www.mimesisinternational.com
info@mimesisinternational.com

MIMESIS EDIZIONI
www.mimesisedizioni.it
mimesis@mimesisedizioni.it

ÉDITIONS MIMÉSIS
www.editionsmimesis.fr
info@editionsmimesis.fr

MIMESIS COMMUNICATION
www.mim-c.net

MIMESIS EU
www.mim-eu.com

www.ingramcontent.com/pod-product-compliance
Lightning Source LLC
LaVergne TN
LVHW100726061126
851857LV00011B/411
* 9 7 8 8 8 6 9 7 7 5 5 6 7 *